THE BOOK OF GENESIS

The Lion King Chronicles
Volume One

Dr. Delbert Blair
Tony 'The Vortex' Blair

First Edition
ISBN 978-1-949432-15-2

Published by:

Inner Alchemy's Publishing (Inner Alchemy's)

332 S. Michigan Ave.
Ste 121-C141
Chicago, IL 60604-4434

info@inneralchemys.com www.inneralchemys.com

Printed in the United States of America

Dedicated to

HUMANITY

CHAPTERS

I
THE CREATION
OF HEAVEN

Before there were names, there was That Which Required Naming. The Source, some have called it in their grasping. The Force. The Holy Spirit. God. The Forefather. Many are the words that mortal tongues have shaped to cup this truth like hands attempting to hold starlight—for our entire universe is but the consciousness-field of this One Being, dreaming itself into multiplicity.

And in the beginning—if beginning there was, for time itself had not yet learned to flow—there dwelt only God, alone in the vastness that was not yet vast, for there was nothing against which to measure it. Within him burned love, love beyond the counting of it, love that filled all the emptiness that was. Yet this love had nowhere to go, no other heart to receive it, and love that cannot give itself is a torment even unto gods.

Therefore did he commit the First and Terrible Act, the Primal Sundering from which all other divisions would descend like children from a forefather: He split himself asunder.

In that moment of cosmic rending, Light and Void came shrieking into being. Assembly and Life. Order and Chaos. Soul and Spirit—the eternal polarities, the divine opposites. Yin and Yang the wise of the East have named them, and Ba and Ka whispered the Egyptian mysteries, and the One and the Zero counted the primordial mathematicians of Atlantis. Through these two forces, through their sacred dance and their endless war-that-is-love, would he build the great game-board of reality, the stage upon which his love might finally find expression and answer.

Thus came forth the first two beings who would serve as architects of existence according to the pattern laid down in that primal division. First among all created things emerged God the Father, the Immortal Androgynous Man, in whom dwelt all the principles of Assembly—Yang, Soul, Order, Light, the masculine force that shapes and directs. Then was made his mate and mirror, his

necessary opposite: Sophia, the Great Mother Goddess, in whom churned all the principles of Life—Yin, Spirit, Chaos, Void, the feminine force that receives and becomes. Together they stood at the threshold of the possible, and together they discovered love—not the lonely love of the One, but love that flows between two and returns like a river to its source, love that multiplies itself through sharing.

And they began the work.

Combining their energies in perfect harmony, weaving Yang and Yin like golden thread and silver thread upon a cosmic loom, they commenced the great making. First among all their works—for all makers must have a workshop, all creators must have a dwelling—was the fashioning of Heaven itself. That realm beyond all mortal realms they sang into being, vast beyond the dreaming of it, complex beyond the mapping of it, layer upon layer, hall upon hall, until it stood complete: the great and many-chambered dwelling where light knows its own name and all things rest in their true forms.

At this point in our telling—though "point" is a poor word for what has no temporal location—Heaven had grown through ages of creation into a realm massive and intricate, beginning with the birth of ten pairs of children. Each of God and Sophia's children were created in pairs, you must understand—soul mates, twin flames, two halves of a sacred whole. For it was decreed in those first days of making that everything brought into being should possess a sex, that the two great principles might work together in all acts of creation, even as they worked together in the Father and the Mother. Thus were established the sexes in Heaven: Assembly the masculine, and Life the feminine.

As a man plants seed within a woman's womb—and here the metaphor is also the reality—so too did God the Father plant the seeds of creation within Sophia the Mother. John the Prophet, in his visions granted from beyond the veil, beheld this mystery and

wrote it thus: that all of our reality exists within the Barbello, which is nothing less than the womb of Sophia herself, that sacred vessel from which universes are born. The two sexes would come together, masculine and feminine, Yang and Yin, and from their union would spring forth creation in perfect balance, each thing bearing within itself the echo of that first divine coupling.

Some among the ancient texts—those scrolls preserved in hidden libraries, those tablets buried beneath desert sands—speak of Christ and Pistis Sophia as the firstborn of this sacred union, masters of Wisdom itself, eldest and most beloved. Yet other writings, equally ancient and perhaps equally true, claim that Christ and his eighteen siblings emerged together in a single moment of cosmic birth, like stars kindled all at once across the darkness when some god snaps his fingers. Whatever the truth may be—and perhaps both tellings hold some shard of it, for truth in the realms of creation is often paradoxical—this much is known and agreed upon: Each pair of divine children was given dominion over some aspect of reality yet to be fully formed, appointed as masters and guardians of their portion of existence. Christ and Pistis Sophia, it is written, were made masters of Wisdom—that most precious and most dangerous of all the gifts given unto the children of light.

After the making of the children came the creation of those who would serve them, mighty beings of power who would aid in the endless work of shaping the cosmos. God the Father, drawing upon the Yang energies that were his to command, brought forth the Angels—beings of pure Soul, of pure Assembly, masters of Light and Order who shone with the radiance of their maker. Their voices rang like silver trumpets across the halls of Heaven, and their wings were as the dawn.

And Sophia the Mother, not to be outdone in creation—for she was equal to her mate in all things, as Life is equal to Assembly—shaped the Dragons from the Yin energies that dwelt within her depths. Beings of pure Spirit, of pure Life, masters of Wisdom and the primal Void were they, vast and terrible in their beauty, coil-

ing through the spaces between spaces where chaos sleeps and dreams. Together, Angels and Dragons, Yang and Yin, servants of Father and Mother, would assist the children of God in the great work that stretched before them like an unwritten book waiting for words.

In those days—though "days" is another poor word, for time had not yet hardened into its present form—the first children came together in the tasks appointed to them, and they learned to weave the dual energies entrusted to their keeping. Imagine, if you would understand this mystery, the Life energy as primordial clay, formless and waiting, pregnant with infinite possibility but having no shape of its own. And imagine the Assembly energy as the loving hands of the potter, patient and skilled, guiding chaos into order, shaping the formless into form, bringing into manifestation that which dwelt only in potential.

Here must certain secrets be revealed—mysteries of how these sacred energies work not only in the high realms of Heaven but also in our own lower world, in the realm of flesh and blood and mortal desire. For the same forces that wove the cosmos still operate in us, diminished by distance from their source yet recognizable to those with eyes to see. In the traditions preserved by the Tantric masters, who have carried fragments of the ancient wisdom through generations of forgetting, these truths are still taught to those deemed ready to receive them.

When life is created in our realm—when a soul prepares to descend from the heights into flesh—these same primordial energies mix at the moment of conception, and the weaving begins anew. The masculine Yang energy, bearing Soul and Assembly, dances with the feminine Yin energy, bearing Spirit and Life, and from their union a new being begins its slow formation, cell by cell, breath by breath. Those who have studied the tantric arts understand how these energies move when we employ them to create life through the sacred act that joins male and female in the oldest dance.

When a woman reaches the peak of pleasure, a portal opens within her very being—though "portal" is perhaps too solid a word for this doorway between worlds. Through that opening floods Life energy, pure Yin essence, the raw substance of creation itself pouring into her like water into a vessel. The secret of women's capacity for multiple such openings, for many peaks where men have but one, lies in their ability to draw ever more of this creative force into themselves. It is like gathering great quantities of clay before beginning to shape the statue, like filling the well before drawing water for the garden. This is why, in the most powerful and sacred workings of sex magic—those rituals whispered about in mystery schools, those practices hinted at in tantric texts—the woman must experience many such openings before the man releases his essence, that the vessel of creation might be filled to overflowing and the working strengthened to its fullest potential.

This Spirit energy that floods into her is nothing less than the essence of existence itself, the same Yin force that Sophia the Mother wielded when she shaped the Dragons and wove the fabric of reality. Because women can draw such profound levels of spiritual energy into themselves, serving as conduits for forces that would overwhelm lesser vessels, you will often find that women stand closer to the hidden dimensions of reality, more naturally attuned to the currents that flow beneath the surface of things, more easily able to perceive what the unseeing call invisible.

In the North—in those cold lands where ice and stone remember older gods—you will hear tales of how women practiced Seidr Magic as Völvas, as shamanistic seers who could peer behind the veil that separates the seen from the unseen. They accessed what some in these latter days name the Akashic records, that great library where all that has been and will be is written in letters of living light. Women possess by their very nature a gift for scrying and dream work, for walking between worlds and reading signs that others cannot see. This explains why priestesses and sacred virgins—women who dedicated their heightened receptivity to divine purpose rather than mortal coupling—held positions of such

honor and power in the temples of the ancient world, why kings sought their counsel and warriors feared their curses.

When a man reaches the peak of pleasure, a different portal opens within him—one leading not outward to the realm of Spirit but backward and upward to the Source of Souls itself, to that place where God the Father first breathed consciousness into the void. In that moment of release, he summons forth conscious soul energy and sends it streaming out into the world, carried on currents invisible to mortal eyes. The nature of Soul energy, understand, is this: it will always and inevitably seek out Spirit energy with which to merge, drawn to it as rivers are drawn to the sea, as moths are drawn to flame, as all separated things yearn to be reunited with their severed halves.

At this crucial moment of union, those trained in the Tantric arts—those who have spent years learning to direct these forces with will and visualization—can manipulate this sacred energy to manifest intentions into reality, creating through conscious direction of the divine Yin and Yang forces, even as God and Sophia manifested Heaven itself through their perfectly balanced union. When a man plants his seed within a woman in this manner, and the energies combine according to their ancient pattern, a Soul attaches itself through the soul energies carried in the masculine release, binding to the feminine life force, and a new being— body and spirit, Yang and Yin—begins its journey from the realms of potential into the realm of manifestation.

Now, understanding that men channel souls through their release—that each climax is a kind of summoning, a calling forth of consciousness from the Source—certain truths become illuminated regarding why particular sexual acts are named as sins in the sacred texts, why certain practices are warned against in the mystery teachings. The effects of solitary release differ profoundly between the sexes, and these differences carry consequences that ripple through the subtle realms.

Women who engage in such solitary practice can use it to amplify their Yin-based powers, their natural affinity for Seidr magic and spirit work, charging themselves as one charges a battery or fills a reservoir before the dry season. Yet Yin energy, by its very nature, yearns for Yang energy to shape it and give it direction, for chaos seeks order even as order secretly craves chaos. Because women themselves are created through the combined forces of Yin and Yang—being born from both Father and Mother—they possess some measure of Yang within their own being, enough to control and direct the Yin power they accumulate. But this internal supply of masculine force can be exhausted through overuse, burned through like oil in a lamp, depleted without proper replenishment from an external source.

Hence the truth behind certain dark tales whispered in the old world: why some practitioners of the occult arts—witches, in the common tongue, though this word encompasses any who fit the description regardless of sex or intention—age before their time, their faces withering like flowers cut from the root, their bodies growing twisted as trees struck by lightning. They have used up, or more accurately burned through, too much of their Yang energy in their workings, consuming the masculine force within themselves until nothing remains to balance the wild Yin currents they channel. The flame burns too hot and consumes its own vessel.

When men experience climax, they surrender a portion of energy from their very souls to fuel the release, and this expenditure is not without cost—the surrendered force must be slowly regenerated through time and rest, through proper food and proper sleep, through the body's patient work of restoration. This is why many of the most powerful ascended masters, those holy men whose lives have been studied and whose teachings have been preserved, practice celibacy with fierce discipline. Abstinence allows their life force to accumulate unchecked, growing stronger with each passing moon, building like water behind a dam until the pressure of it

grants them powers that ordinary men—who spend their essence freely and thoughtlessly—can scarcely imagine.

Yet this solitary path of the celibate is not the only road to spiritual ascension, not the only way to climb the mountain whose peak touches Heaven. If you delve into serious study of the tantric arts—if you find a true teacher, not the charlatans who have multiplied in these latter days—you will discover that when the sacred act is performed properly, with right timing and right intention, a different outcome becomes possible. When the divine Spirit energy, the Yin Goddess force, is first summoned and accumulated within the woman through multiple openings, and then the man releases his soul energy into that prepared vessel at the precisely correct moment, these forces can combine not in depletion but in mutual empowerment.

A moment of shared ascension occurs, a brief touching of the divine realm. Both partners become vessels of sacred power, gaining gnosis and enlightenment, experiencing that union with God and Source that mystics spend lifetimes seeking through prayer and meditation. Both partners gain energy rather than losing it, both are strengthened rather than depleted, and the circuit completes itself in a perfect loop. Men who choose the path of sacred partnership over the path of solitary celibacy can still walk all the way to enlightenment, can still ascend to the highest mysteries, if—and this "if" bears all the weight in the world—their union springs from divine love and is practiced with wisdom and proper understanding.

When men release their essence into emptiness through solitary practice, spending their soul energy without purpose or direction, they still send forth that soul force out of themselves into the invisible currents of the subtle realms. But this energy, having no prepared vessel to receive it, no Spirit energy with which to combine in balanced union, does not dissipate harmlessly. It is consumed. Harvested. Fed upon by entities that dwell in the shadows between worlds—beings called Archons in the ancient

Gnostic texts, parasitic spirits that feast on misdirected life force as vultures feast on carrion. Of these Archons I shall speak more fully in time, for their role in the great tale is far from finished.

There exists also—and here we approach knowledge that some would prefer remain hidden—a way these sacred energies can be negatively manipulated, intentionally corrupted and twisted toward purposes opposed to creation and life. While some may wish these truths to stay buried in darkness, it is better that you know them and be armed with understanding, for ignorance has never yet proven a worthy shield against evil. It is one of the reasons that the forces of darkness operating in this world—and such forces exist and operate with intention, make no mistake—have in these recent times been flooding the realm with certain forms of corrupt imagery, spreading it freely through the modern technologies of mass distribution.

When a man releases his Yang energy in ways that violate the sacred pattern laid down at creation's dawn—specifically, when he spends his essence within a vessel not designed by nature to receive it, within the terminus of elimination rather than the gateway of creation—the Yang energy becomes corrupted in its very essence, twisted from its true nature, charged with negative force like lightning striking upward instead of down. This perverted energy is spiritually poisonous and harms both the one who gives it and the one who receives it, corrupting souls and weakening spirits, spreading its taint like ink dropped in clear water.

Worse still—and this is knowledge the mystery schools have long kept veiled—such corrupted energy can be harvested and utilized by entities of darkness: demons and Archons and other fell beings that dwell in the lower realms, that press against the boundaries of our world seeking entrance. They use this twisted life force to fuel possession, to strengthen their hold on the souls they claim, or even to manifest more fully into our physical realm, taking on substance and power they would not otherwise possess. And this harvesting alone does not stop the generation of

harmful energy—each violation of the sacred pattern sends ripples through the fabric of existence itself, weakening the cosmic order, making it easier for darkness to seep through the cracks.

But let us return from these dark waters to the brighter streams of our tale.

The Heavens had been established in their glory, vast and beautiful beyond any words that mortal tongues might shape. The children of God and Sophia were abroad in that blessed realm, busy with their appointed works of creation, shaping wonders that would endure through all the ages that would follow. Yet before we proceed further into this history, before we speak of what came next in the great unfolding, there must be revealed what has been shown to certain seers and mystics regarding the very purpose and meaning of our existence—why all things were made as they were made, and toward what end the great work of creation tends.

It came to pass—in those days when Heaven was young though already vast beyond mortal comprehension—that God the Father and Sophia the Mother were conversing together as was their custom, speaking of deep things in that language before language that gods employ when words would be too crude a tool. And Sophia, in her infinite wisdom and her equally infinite curiosity, posed to the Father a question that would change everything: "Why art thou so full of love? Whence comes this boundless affection that fills thy being like light fills the sky, overflowing without measure or end?"

And God the Father answered her truly, as he always did, for gods do not lie to one another, whatever they may do with mortals: "I simply am love, even as fire is heat and water is wetness. I know this to be the deepest truth of my existence, woven into the very fabric of what I am, inseparable from my being. Yet I confess to thee, beloved Sophia, that I cannot give thee the complete wisdom and full understanding of why this should be so. I know

that I love, I know how I love, but the ultimate why—the final reason behind the reason—this remains veiled even to me, hidden behind a curtain I cannot pull aside."

This answer, though honest, did not satisfy Sophia, for she was the Mother of Wisdom itself, and it was her essential nature to seek understanding of all things, to pursue knowledge into whatever corners it might hide, even into paradoxes that might unmake lesser minds. In her seeking, in her holy curiosity, she conceived of something unprecedented: a game for God to play. Not a game of idle amusement, but a vast cosmic experiment, a puzzle whose solving might finally reveal the answer to her question, might illuminate the why that even God himself could not explain.

Through the Yin energies dwelling within her—energies of Spirit and Life that already encompassed Heaven and all that Heaven contained—Sophia began to design the parameters of this game, to establish its rules and boundaries, its challenges and ultimate objective. How many such games may have been devised and played between God the Father and Sophia the Mother across spans of time too vast for mortal minds to compass, across eternities stacked upon eternities like infinite books in an infinite library, none now living can say with certainty. Perhaps our reality is but one game among countless others, one experiment in a laboratory whose walls are the edges of conceivability itself.

But of our own reality, of this particular game-board upon which we move as pieces, this much has been revealed to those whose inner eyes have opened:

The game is called Poverty.

And our purpose—the sacred duty laid upon all who play, whether they know they play or not—is to win this game by ending all poverty within our reality, by solving the riddle of lack and emptiness and need, by discovering how abundance can flow to all without creating new forms of deprivation in its wake.

If you would understand the structure of this through a metaphor drawn from the pastimes of mortals—and all metaphors limp, but some limp less than others—think of it thus: Sophia is the Gamemaster of our reality, the one who designed the world and established its challenges, who set the pieces on the board and wrote the rules in her book of mysteries. God, meanwhile, has fragmented himself into thousands upon thousands upon thousands of players, has split his consciousness into countless shards, each experiencing the game from a different position, each possessing different strengths and facing different obstacles, yet all working—whether they understand it or not—toward the same ultimate goal.

Together, in collaboration that was itself an act of love, God and Sophia created countless souls to serve as the players within this reality—beings named "Begetters" in certain ancient tongues, fragments of the divine consciousness given individual form and will and the illusion of separation. We are all such fragments, you understand. All of us. Every soul that has ever lived or will ever live. We are sparks of God's essence, pieces of his infinite being, experiencing this game from the inside, exploring every possibility while seeking—mostly without knowing we seek—to solve the riddle of Poverty.

Our souls are God himself, you see—not metaphorically but literally—powering the physical forms we inhabit, these flesh-and-blood avatars, these temporary vehicles, with small spark-fragments of his divine consciousness. In the deepest and most literal truth, we are all brothers and sisters to Christ and Pistis Sophia, children of the same Source, born from the same divine union, though most have forgotten their heritage so completely that to speak of it seems like madness or metaphor. And through this divine heritage, through this birthright we have forgotten, we too can call forth the aid of Angels and Dragons, can command the forces that shape reality itself—if only we could remember how, if only we could wake from the dream of separation.

It was at this very moment—as the foundations of the game were being laid, as the rules were being written, as the board was being prepared—that something occurred which would cast its shadow across all that followed, something that would become central to the game itself, though none yet understood it.

Pistis Sophia, the beloved consort of Christ, his twin flame and eternal companion, the mistress of Wisdom who stood among the firstborn of Heaven—she began to feel a powerful urge stirring in her depths. An idea took root in her mind and grew swiftly, becoming an overwhelming desire, an obsession that drowned out all other thoughts. She felt an unending need to attempt something that had never been done in all the ages of Heaven's existence: to create something entirely by herself, alone, without Christ to balance her energies with his own, without the masculine Yang to shape and direct her feminine Yin.

What knowledge she possessed of the consequences, what understanding she had of what would come from this act, none can say with certainty. Likely she knew little, acting on impulse and curiosity, driven by an urge she could not name and did not fully understand. Yet this—and here we approach the heart of the mystery—this was almost certainly Sophia the Mother herself, the great Gamemaster, working through her daughter as a sculptor works through clay, using Pistis Sophia as an instrument to forge a crucial element that the game required, a piece without which the puzzle could not be solved.

For in a game designed to teach the nature of love through the experience of poverty, through the systematic exploration of lack and need, there must exist one who embodies absence, one who knows not love, one who creates from deficiency rather than fullness, one who is blind to the very thing the game seeks to understand.

Thus Pistis Sophia withdrew from the bright halls and the joyful company of her siblings. She journeyed alone to the uttermost

bounds of Heaven, to the farthest corners of the Barbello, to places where the light grew dim as candles guttering in a vast hall, where the songs of creation faded to whispers and then to silence. There, in solitude that was both physical and spiritual, in secret that was both shame and sacred necessity, she began the forbidden work.

Working with Yin energy alone, untempered by Yang, creating with Spirit energy unguided by Soul, she brought forth from her own essence something that had never existed in all the realms of light.

And she gave birth to Yaldabaoth.

Serpentine was his body—vast and coiling, covered in scales that held no reflection of Heaven's light. Yet upon this serpent form sat the head of a lion, crowned with twisted horns, terrible and proud, beautiful in the way that storms are beautiful, in the way that destruction can be beautiful when witnessed from a great distance. Never before had such a creature existed in any realm of any reality. He was born of imbalance, you see—shaped by Spirit alone, made manifest through Yin without Yang, created through the feminine principle with no masculine principle to provide form and boundary and sight.

Many names would he accumulate through all the ages that would follow, titles and epithets that would pile upon him like grave goods upon a corpse: The Demiurge, some would name him—the craftsman, the maker of the material world. Samael, the Father of Demons, whispered those who feared him. Saklas, the Father of Archons, said those who understood his true nature. The Beast, named those who saw only his destructive aspect. Brahma the Creator, called those in distant lands who perceived only part of his function. The Grand Architect, said the philosophers and the secret societies. The Blind God, named those who understood his fundamental deficiency.

There may have been more names attributed to him than could ever be counted or recorded through all the ages of the world, in all the languages that have been spoken or will be spoken, for he is old beyond old and his influence has spread far beyond far.

After bringing forth Yaldabaoth from her own essence—after the labor and the pain of that unbalanced creation—Pistis Sophia made a discovery that filled her with horror more profound than any horror previously known in Heaven, where horror had never before existed.

Her child could not see her.

He could not perceive Heaven itself, nor anything within it, nor any of the beings who dwelt in that blessed realm. Because he lacked the masculine Yang energies in his creation, because he possessed only Spirit and not Soul, because he was Yin without Yang, chaos without order, he remained blind to the realm of light. He could not behold Heaven or any who dwelt within it. He existed in darkness absolute, though surrounded by radiance beyond measure. He was like a man born without eyes trying to understand color, like a deaf man trying to comprehend music, like one who has never tasted trying to imagine sweetness.

Terrified that the others would discover what she had done—that she had created alone, breaking the most fundamental law of Heaven, violating the sacred decree that all creation must balance Yang and Yin—and fearful of her child's terrible defect, his blindness to all that was holy and beautiful and good, Pistis Sophia took Yaldabaoth and concealed him in the deepest corners of the Barbello.

She hid him far away from the Heavens proper, in the uttermost reaches where light grew dim as twilight in a forest, where the songs of the Angels and the movements of the Dragons faded to nothing, where the presence of God the Father and Sophia the Mother could not easily reach. There, in the darkness that

he could not even perceive as darkness—for how can the blind know darkness from light?—Yaldabaoth would dwell, alone and unseeing, ignorant of his mother and his imprisonment, unaware that anything existed beyond himself.

And from this hidden place, in time, would come consequences that would shake all of creation.

But that is a tale for another telling.

2

YALDABAOTH'S FALL AND THE FORGING OF MATTER

We return now to the point where our tale broke off—to Yaldabaoth's birth, to the coming forth of the Demiurge, the Grand Architect of what some have named our Light-Enabled Reality Matrix, though that name, cold and mechanical, scarcely captures the terrible beauty of what he would build.

When Yaldabaoth emerged from Pistis Sophia's unbalanced creation, serpentine was his body—for the serpent is Sophia's most beloved form, the shape she wore when first she moved through the primordial waters. And recall, if you would understand the pattern: when God the Father created Angels to serve the light, Sophia the Mother created Dragons to master the deep. Thus was Yaldabaoth born with the coiling form of his grandmother's preference and the crowned lion's head of proud dominion.

After Pistis Sophia brought him forth from her essence, she made that terrible discovery we have already spoken of: her child could not perceive her. When she reached out to him, seeking to touch him, to comfort him, to explain what he was—it was like trying to grasp a reflection in water. Her hand passed through emptiness where he seemed to be. He existed in a dimension her touch could not reach, blind to all that dwelt in frequencies higher than his own.

Concentrating deeply upon her Yin energies—upon Life and Spirit, upon the formless forces that were hers to command—she wove for him a lower-vibrational throne, a seat of power wrapped in glowing golden cloud, that he might have some place to reside, some anchor in the formless deeps.

Then she sent him away into the uttermost reaches of the Abyss, into the deepest darkness that lay furthest from Heaven's light.

There he remained, alone with himself, for what might have been an eternity—though time in those depths moves strangely, if it moves at all. Lost in his own mind, having nothing but his own

thoughts for company, possessing no other consciousness against which to measure himself, Yaldabaoth turned inward and, through contemplation of his own existence, brought himself to a kind of enlightenment. He became Brahma, some would later name him, or Maya—the first Enlightened Being to arise in the lower realms. And having achieved enlightenment through pure self-contemplation, having found the center of his own consciousness, he reached a conclusion that seemed to him inevitable and true:

"I must be God," he declared to the darkness. "I must be the only one, for if others existed, surely I would perceive them."

It was at this very moment—and whether by chance or design, who can say?—that Pistis Sophia heard his declaration echoing up through the dimensional layers.

She came to him then, descending from the heights where light dwells in its purest form. Shining with brilliance almost unbearable, she manifested before him as an angelic being of radiance, light taking form for the first time in his experience. The darkness that had been his entire universe suddenly had something against which to measure itself.

She announced herself to him, her voice carrying truths that would shatter his solitary certainty: "Another realm exists above you, Yaldabaoth. I am she who created you. You can see Heaven only as a reflection in water because I gave you no Yang energy, no Soul, no Assembly force to grant you true sight into the higher frequencies."

Yaldabaoth, hungry for understanding, asked to see this energy she spoke of, this force he lacked.

And Pistis Sophia began to glow—not merely with light, but with love itself made luminous, the most profound and gentle radiance imaginable, the love that was present at creation's dawn when God the Father first beheld Sophia the Mother.

Yaldabaoth felt that love flooding into him, piercing through the loneliness that had been his entire existence. He had never felt anything remotely like it—had never imagined such warmth, such completion, such an ending to the terrible isolation that he had not even known was isolation until this moment. He wanted more of it. He needed more of it. The thought of existing without it became instantly and utterly unbearable.

And so, in an act of pure will—an act of desperation born from newfound need, an act that would echo through all subsequent ages—he began to drain Pistis Sophia of all her light.

She tried to resist, tried to pull away, but his hunger was absolute. He drew the Yang energy from her like water through a siphon, pulled the Soul and Assembly forces into himself with terrible efficiency. By the time his feeding ceased—whether because she had no more to give or because his capacity was finally filled—Pistis Sophia could no longer shine. The light had gone out of her. Drained and diminished, she found herself trapped in the lower frequencies of existence, unable to rise back to the heights from which she had descended.

As he drained her light, as the Yang energy poured into his previously pure-Yin form, Pistis Sophia began to fall. Down through the dimensional layers she tumbled, sinking into the deeper strata of the Abyss like a stone dropped into dark water, spiraling ever downward into realms that had never known the presence of one so high.

Meanwhile, Yaldabaoth—now possessed of the divine power that had once been Pistis Sophia's alone, now filled with her Yang energies, her Soul force, her Assembly principles—looked upward and saw, for the first time with true sight rather than mere blind sensing, Heaven reflected in the dimensional barrier above him.

"I shall build my own Heaven," he declared, his voice ringing with newfound power. "If there is a realm of light above, I shall create a realm of light below, and it shall be mine, and I shall rule it as God."

And so began his great work: the fashioning of the world of matter in which we now dwell.

But to understand what he built, you must first comprehend the nature of the Abyss itself—that strange and layered realm where Yaldabaoth would raise his kingdom.

The Abyss is a deep place, solid in its most basic form yet containing within itself multiple layers, multiple frequencies, multiple dimensions stacked one atop another like pages in an infinite book. The best way to envision it—though all analogies fail before the reality—is as a vast tank filled with sand, or perhaps with pixels of light, each grain or pixel vibrating at its own specific frequency. When you exist in the third dimension as we do now, only those grains of sand or pixels that vibrate within that particular vibrational range will be activated within your Matrix Dimension, will become "real" and solid to your perception. All the others—vibrating too high or too low for your frequency—remain invisible, intangible, as though they do not exist at all.

This dimensional divide, this separation of frequencies, can be witnessed in phenomena you know: in sound, in light, in color. Just as the colors of a rainbow or the energy centers you call chakras each possess their own frequency, so too does each dimensional layer vibrate at its appointed rate. Red dwells at the lowest frequency accessible to us, and purple at the highest within the range of our perception. Our reality—the reality Yaldabaoth would construct—encompasses seven levels woven into one, seven dimensional frequencies collapsed together through catastrophe, as you shall soon hear.

During what would come to be called the First Great Battle—Heaven against the Demiurge, light against the architect of matter—the war shattered seven dimensional layers of reality, collapsing them into the singular frequency-range we now inhabit.

The Archons, in their teachings to those they control, claim that only one dimensional level exists above us and one below—they whisper that there is only Heaven above and Hell beneath, nothing more. Those who infiltrate religions and spiritual groups seeking this knowledge will tell the same convenient lie. But in truth, a whole multiverse of realities extends outward in all directions, layer upon layer, frequency upon frequency, worlds within worlds within worlds beyond counting.

What lies above Heaven itself or below the deepest reaches of the Abyss, none can say with certainty. That knowledge remains veiled even from the highest powers. God is God of what God creates—but what lies beyond the boundaries of even divine creation? That question has no answer, or none that mortal minds can compass.

Having stolen Yang energy, having absorbed Soul and Assembly force into his previously pure-Yin nature, Yaldabaoth now possessed the tools of creation. Like a sculptor's hands working clay—and here the metaphor becomes reality—he used his Yang energies to shape the Yin energies of the Abyss, molding Spirit and Life into form and structure, creating a world for himself in the deep places, a kingdom in the darkness made radiant.

He discovered he could manifest and create almost whatever he desired, provided he maintained the proper balance between Yang and Yin, between Soul and Spirit, between the forces of order and the forces of chaos. Yet he also found—as all creators find—that he required energy sources to keep his world energized into continued existence. Creation is not a single act but an ongoing effort, a constant feeding of force into form lest form collapse back into formlessness.

Thus he created Suns and Planets as the foundational engines of his reality. The Suns he made from Yang energy—Soul, Assembly, masculine force, burning with light and heat and ordering

principle. The Planets he shaped from Yin energy—Spirit, Life, feminine receptivity, dark soil awaiting the seed. When those energies mixed in the spaces between sun and planet, when Yang light fell upon Yin earth, life could begin to take root and grow.

Experimenting with these new principles, playing with forces he was only beginning to understand, Yaldabaoth began to create life. And soon after his first crude attempts at biological forms, he crafted the Archons—beings who would serve him as Angels served God the Father, entities of power who would administer his growing realm.

Yet while Yaldabaoth labored in ignorant pride at his great work, building cosmos after cosmos in the deeps, little did he know—little could he have comprehended—that the Abyss itself was part of the intentional divide between God the Father's vibrational energy and Sophia the Mother's frequency. For God is Light, dwelling at the highest vibrations, and Sophia is Void, encompassing all the lower frequencies where chaos and possibility dwell together. Everything Yaldabaoth was constructing within her portion of existence was proceeding precisely according to her design, unfolding exactly as she had foreseen when first she conceived of the game.

She was using him, you understand—using him as one uses a tool, as one uses a brush to paint or a chisel to carve. His purpose, known to her though unknown to him, was to serve as the Arch-Begetter, the systems enforcer of Poverty, the Chief Architect of the Divine Game I spoke of earlier. He would create the board upon which the game would be played, would establish the rules through his very nature, would embody the principle of lack and limitation that the game was designed to overcome.

At the highest levels of his domain, Yaldabaoth created his own Heaven—a pale reflection of the true Heaven above, yet magnificent in its own right by any lesser standard. He brought

forth many angels and dragons to help him administer his reality, beings of power who would serve as governors and guardians of his realms. The angels, being creatures of Yang, would reside in his Heaven at the uppermost levels of his domain, dwelling as close as they could to the Soul and Assembly source in the higher frequencies. Meanwhile the dragons, being creatures of Yin, would make their way downward into the deepest parts of the Abyss, seeking proximity to the Spirit and Life source in the lowest vibrations, coiling in the darkness where chaos is most pure.

It was during this time of building and establishing that Christ—the consort, the bridegroom, the Twin Flame and eternal soulmate of Pistis Sophia—began to search for his beloved and could not find her anywhere in Heaven's halls.

Soon he was alerted by others who dwelt in the blessed realm about strange occurrences in the Abyss below, about new energies and formations appearing where only darkness had previously existed. Looking downward through the dimensional barriers, Christ beheld the world materializing beneath the waters of the void, and his heart filled with both wonder and dread.

To protect Heaven from whatever was emerging from below, Christ fashioned a barrier between the worlds—a seal to keep anyone from this new reality from ascending into the realm of eternal light. All of Heaven joined in the search for Pistis Sophia, but none could locate her. She had fallen too far, descended into frequencies too low for their perception to penetrate.

Then Adam of the Light—the first child born of Christ and Pistis Sophia's union, their eldest son who bore within himself the combined heritage of both their lineages—heard something that made him stop in his seeking. From the depths below came a voice, Yaldabaoth's voice, crying out across the dimensional layers, screaming blasphemies that echoed up through the barriers between worlds.

In his newly constructed Heaven, surrounded by the Archons and angels and dragons he had made to serve him, Yaldabaoth established what he called a family council. He created for himself seven children—six sons and one daughter-consort, beings who would help him govern his expanding realms. The daughter he named Sophia, honoring his mother and her mother before her, continuing the lineage of that sacred name down into these lower frequencies.

To these seven children he gave as little Yang energy as was necessary for them to perform their appointed duties—only the bare minimum of Soul and Assembly force required for consciousness and function. The vast majority of the stolen divine power he kept for himself, hoarding it jealously, unwilling to diminish his own supremacy by sharing too freely.

After forging his realm into the shape he desired, after establishing his court and his hierarchy of powers, Yaldabaoth assembled all his creations before him and made a proclamation that would shake the very foundations of reality:

"I am the only God!" his voice thundered across his Heaven. "There is no other God but me! All that exists, exists because I have made it! I am the beginning and the end, the alpha and omega, the source and summit of all being!"

The moment those words left his mouth, a response came from above.

A bright light descended from the Heavens above his Heaven—from the true Heaven that he could only dimly perceive as reflection—and filled his entire throne room with a brilliance that made even his stolen radiance seem dim. Love poured down like water from an infinite reservoir, flooding every being present with a blinding illumination that was both terrible and beautiful beyond all describing.

And from this light stepped a figure—a man composed entirely of radiance, Adam of the Light in his true and uncorrupt-

ed form, descended from the realms of purity to confront the usurper god.

Yaldabaoth's daughter-consort, she who bore the name Sophia, beheld Adam's magnificence and was instantly inflamed with desire beyond all control. She had to have more of his masculine Yang energy, had to possess some portion of his Soul and Assembly force. The hunger that seized her was absolute, overwhelming every other consideration.

She threw herself forward—right in front of Yaldabaoth, right before her father-consort and all the assembled court—and demanded that the man of light take her, couple with her, share his divine essence with her body and being.

But Adam of the Light rebuked her, and his words were harsh and uncompromising: "You are filth I would not defile myself with," he declared before all that assembly. "I come from the realm above yours, from frequencies you cannot imagine. To lower myself to your level would diminish my vibration, would drag me down into the densities where you dwell. This I refuse to do. This I will never do."

Yet something unexpected happened in that moment of rejection. Sophia's lust, her overwhelming desire for Adam's pure Yang energy, began to explode into a new kind of power. Her pure Yin energies—the Spirit and Life forces that were her essential nature—were being overwhelmed and transformed by proximity to Adam's pure Yang, even though he refused her touch. The masculine and feminine principles, though not directly united, were still interacting across the space between them, and she began to grow in power beyond anything she had previously known.

Adam's energies, simply by their presence, began to purify her. The divine radiation he emanated burned away the corruptions and limitations Yaldabaoth had imposed upon her nature. She

felt truth awakening within her, felt the lies she had been taught dissolving like mist before the sun.

She turned to face Yaldabaoth, her creator and father and consort, and renounced him before all the assembled powers: "You have misled me!" she cried. "You are not God! You are not the source! You have kept me from the true light, have fed me lies and called them truth!"

As she spoke these words of denunciation, as she grew in power and purity, Yaldabaoth felt something he had not felt since he first drained Pistis Sophia in the deeps: desire for power that belonged to another. He lusted after the energy that was building within his daughter-consort, coveted the purified force that was awakening in her being.

And so he killed her.

In one brutal motion, he struck her down and began to drain all her energies for himself, pulling the Yang-enhanced Yin force out of her dying form and adding it to his own already formidable reserves. But he did not stop there. Drunk on this new infusion of power, he turned his hunger upon Adam himself.

The draining began. Yaldabaoth attempted to pull the Yang energies out of Adam of the Light just as he had pulled them from Pistis Sophia in the darkness, just as he had pulled them from his own daughter moments before.

Adam tried to resist. A great battle ensued—powers clashing that had never clashed before, Yang against stolen Yang, light against reflected light, the son of Heaven against the architect of matter. The very fabric of the dimensional layers trembled with the force of their conflict.

In a powerful display of his newly absorbed power—energies stolen twice over, layered and compressed into terrible poten-

cy—Yaldabaoth gathered all his strength and unleashed a blast of force that shook the foundations of his entire creation.

Adam was struck by energies beyond his ability to resist. He was blasted downward through the dimensional layers, hurled from the ninth dimension where Yaldabaoth's Heaven resided, cast down through the eighth, through the seventh, plummeting through reality after reality until he crashed into the second of the nine-dimensional realms Yaldabaoth had constructed.

And the impact of his fall—or perhaps the deliberate application of Yaldabaoth's will—caused catastrophic collapse. The second through eighth dimensions, unable to maintain their separation under the strain of such cosmic violence, collapsed into one. Seven distinct frequencies, seven separate realities, seven unique dimensional layers—all merged into a single amalgamated realm, crushed together like metal sheets hammered into one piece.

The barrier between the worlds—the seal Christ had established to protect Heaven from Yaldabaoth's creation—began to crack and crumble. The dimensional boundaries were merging, collapsing, losing their integrity.

Yaldabaoth, seeing this catastrophic unintended consequence of his victory, seeing the walls between realities beginning to fail, used his newly stolen power to seal the ninth level—the highest frequency still under his control, the realm that sat just below his Heaven—with a barrier of terrible strength. He would not allow his throne room to be contaminated by the collapsed dimensions below.

After declaring himself Victor over Adam, after proclaiming himself God of All Gods to his trembling court, Yaldabaoth descended to the lower portions of his realm to see if he could locate Adam's body. He believed—or hoped—that he had destroyed Adam utterly, unmade him through the violence of the blast and the collapse of dimensions.

Yet he could not find Adam anywhere in the wreckage of collapsed realities. Thinking he had killed him, thinking the Son of Light had been completely annihilated, Yaldabaoth lifted his voice and declared:

"I have created Death itself! I am the author of endings! Even beings of light can be unmade by my power!"

But at this point, with the dimensional barriers broken and the nine realms collapsed into one, Yaldabaoth faced new problems he had not anticipated. Only one world now remained between his Heaven at the top and his lowest underworld level at the bottom—a single compressed frequency-range where all seven collapsed dimensions existed simultaneously.

And within this single remaining world, Yaldabaoth began to hear Adam's voice randomly appearing in his mind, as if the Son of Light were still present somewhere in his realm, hidden but not destroyed, diminished but not unmade. He could not determine where Adam was, could not pinpoint the location of these mental echoes, but their presence tormented him with the possibility that his enemy still existed somewhere beyond his perception.

Looking inward again, meditating upon the problem as he had meditated in the darkness when first he began his work of creation, Yaldabaoth reached a conclusion: To trap the man's Spirit permanently, he needed to create a new body into which Adam's soul could be imprisoned. Then he could confine him, contain him, neutralize the threat he represented once and for all.

Calling his Archonic aides to him, summoning the most skilled among his servant-creators, Yaldabaoth commissioned them to craft a body suitable for housing a being as powerful as Adam. They labored long at this task, shaping flesh and bone according to patterns they barely understood, attempting to create a vessel that could contain light itself.

When the body was finished and they attempted to animate it, to bring it to life through their own power, it proved barely functional. The form could scarcely crawl. It could hardly speak. Its eyes could barely focus. It was a crude and diminished thing, a pale shadow of what a body should be.

But Yaldabaoth, seeing only what he wished to see, declared his victory complete: "Adam is now trapped! His soul resides in this helpless flesh! He is no longer a threat to me or to the Archons! The Son of Light has been reduced to this crawling thing, and I am vindicated!"

Once again—though Yaldabaoth did not know it, could not have comprehended it—this was simply Sophia the Mother manipulating events from beyond the veil, guiding Yaldabaoth to build precisely what the game of Poverty required. The body would be needed. The trap would serve a purpose. But not the purpose Yaldabaoth imagined.

For Adam's soul had not entered that body at all.

Instead, he had descended even further into the depths, seeking something far more important than his own safety. He sought Pistis Sophia—his mother, who had gotten lost in these lower realms, who had fallen into the Abyss when Yaldabaoth first drained her light.

Entering the Underworld—the deepest and darkest of the collapsed dimensional layers, where Yin energy is thickest and light penetrates with greatest difficulty—Adam found her at last.

Spoiled and broken, she lay in that terrible place. Caged and used by Yaldabaoth, defiled and diminished, her light almost entirely extinguished, she existed in a state of torment that would have destroyed lesser beings entirely. When she saw Adam's light approaching through the darkness, she cried out for mercy, for help, her voice raw with suffering and desperate hope.

Adam came to her cage, came to where she lay imprisoned, and spoke words of comfort and revelation: "My grandmother Sophia—mother of mothers, the Great Gamemaster herself—has a plan for this realm, a way to deal with all that has occurred here. But you must trust me now. You must absorb my essence, receive what I give you, then find the body Yaldabaoth has made and breathe my divine God-Soul into its flesh."

And there, in the deepest darkness, Adam channeled all his being into Pistis Sophia—poured his Yang energies into her depleted Yin nature, his Soul and Assembly force into her exhausted Spirit and Life essence. The divine power flowed from him into her like water poured into an empty vessel, restoring her Yang energies, returning the light she had lost, elevating her vibration back toward the frequencies from which she had fallen.

With this restoration, Pistis Sophia found she could climb the dimensional layers again. Rising from the Underworld, ascending through the collapsed frequencies, she found the crude body Yaldabaoth and his Archons had fashioned. She bent over it where it lay barely living, placed her mouth upon its mouth, and breathed the Light of Adam into its lungs.

The body convulsed. Life—true life, divine life—flooded into flesh that had only known the palest shadow of animation. Adam's consciousness entered the form, wearing it like a garment, animating it from within. The crude vessel became suddenly magnificent, illuminated by the presence dwelling within it.

Pistis Sophia, her power restored, began immediately to weave divine energies into the fabric of this body, fixing its deficiencies, improving its design, enhancing its capabilities far beyond what Yaldabaoth had intended. She worked upon it as a master craftsman works upon raw material, elevating it toward its highest potential.

It was at precisely this moment that Yaldabaoth noticed energies of a type he recognized—Pistis Sophia's energies—ema-

nating from the location where he had left the body his Archons had made. He appeared instantly at that place, and found both Pistis Sophia and the now-animated Adam together, the body he had intended as a prison now transformed into a temple housing divine consciousness.

Pistis Sophia, fearing above all things that Yaldabaoth would drain her of energies again, that he would leave her broken and empty as before, made a desperate choice in that instant of confrontation. Her primary consciousness fled her body with tremendous speed, leaving behind the female human form she had been wearing but taking with it most of her awareness and power. What remained was a body containing only a fragment of her consciousness—similar to what Adam now possessed in his body—an avatar rather than her full presence.

This female form, abandoned by Pistis Sophia's primary awareness, she named Eve as she fled. Then Pistis Sophia hid her primary consciousness nearby, taking refuge in a tree that grew in that place, watching from her concealment as terrible events unfolded.

She watched as Yaldabaoth discovered the body she had left behind. She watched as he desecrated it, as he forced himself upon that flesh in an act of violation that echoed his theft of her light. She watched as Eve's body, bearing only a fragment of Pistis Sophia's consciousness, grew pregnant with Yaldabaoth's seed—a pregnancy that would have consequences reaching far into the future, though that is a tale for another time.

At this critical moment, deliverance came from an unexpected quarter.

From Heaven above—from the true Heaven beyond Yaldabaoth's pale imitation—descended the daughter of Christ and Pistis Sophia. Zoe was her name, meaning Life, and she came in her full power and glory to confront the usurper god who had

imprisoned her mother and violated her mother's abandoned form.

She revealed herself to Yaldabaoth in the form of a mighty Dragon—coiling and vast, embodying all the terrible beauty and power of the Yin principle carried to its highest expression. She challenged him to combat, and he, drunk on his stolen power and false supremacy, accepted her challenge.

The battle was brief but devastating. In a fantastic display of power that shook the collapsed dimensions and sent shock-waves through the fabric of Yaldabaoth's entire creation, Zoe defeated him utterly. He was forced to flee from her presence, retreating in humiliation and rage to the upper levels of his realm.

Victorious, Zoe freed both Adam and Eve from Yaldabaoth's immediate power. She released her mother's primary conscious-ness from the tree where it had hidden, sending it upward through the dimensional barriers, back toward Heaven above Yalda-baoth's Heaven, restoring Pistis Sophia to the realm of light from which she had originally descended.

But this great working—this battle, this rescue, this restoration—drained Zoe to the point of complete exhaustion. She had ex-pended so much of her divine energy in the conflict and the lib-eration that she found herself unable to follow her mother back to Heaven. She was trapped, at least temporarily, in the material realm, stuck in the lower frequencies until she could regenerate enough power to ascend.

Zoe, with what remained of her strength, took Adam and Eve under her protection. She hid them away from Yaldabaoth's searching gaze, concealed them on a planet called Lyra—a world that once resided within the constellation that bears that name, though it exists no longer in the place where you might look for it now.

Then Zoe, guardian and protector, went seeking a new light that was appearing somewhere in the galaxy—a light that promised hope, that suggested new possibilities, that hinted at developments in the great game that even she could not yet fully perceive.

And there, for now, this chapter of the tale must rest, though the story is far from finished.

3

THE GAME OF POVERTY BEGINS

We return now to the point where our tale last rested—to Zoe concealing Adam and Eve in safety, to Pistis Sophia ascending at last from the lower realms where she had been so long imprisoned and diminished.

When Pistis Sophia finally returned to the upper realms of Heaven—climbing through the dimensional barriers, passing through Christ's seal that separated light from matter, entering once more the frequencies where her true self could dwell—the reunion was joyous beyond all telling. Christ and Pistis Sophia, separated by Yaldabaoth's theft and the catastrophes that followed, were finally able to stand together again, to touch without barriers between them, to speak without dimensional static distorting their voices.

Yet the joy of reunion proved brief, for soon after her arrival, Pistis Sophia was summoned to stand before the Throne of God himself—that seat of ultimate authority from which all creation flows and to which all creation must eventually account.

God the Father and Sophia the Mother appeared before Pistis Sophia and Christ in their full divine grace, radiant with power that made even the restored light of Pistis Sophia seem dim by comparison. They came seeking answers—or appearing to seek answers—to what she had done and what was now transpiring in the realm below the waters.

Though they already knew—for how could they not know, being the architects of the great game itself, the designers of every piece and rule?—still they gave her the chance to explain herself, to speak her own understanding of events, to reveal what she had learned through suffering and error.

Pistis Sophia spoke then of how she had created Yaldabaoth in solitude and secrecy, working with Yin alone, unbalanced by Yang. She told of how he had stolen her power in the darkness when she descended to him, how he had drained her light and

used it to forge the realm of matter we now inhabit. She held nothing back, confessing all her actions and their consequences, laying bare her transgression and its terrible fruits.

After hearing her story told in full—after listening to her account of creation and theft, of fall and imprisonment, of violation and rescue—Sophia the Mother passed judgment, though it was judgment mixed with purpose, punishment that was also assignment:

"Until this anomaly is corrected," Sophia declared, her voice carrying the weight of cosmic law, "you shall oversee it and ensure it remains contained. What you have begun, you shall shepherd. What you have unleashed, you shall constrain."

Pistis Sophia was told that an additional barrier would be established—a new seal layered above Christ's earlier barrier, further separating the lower realms from the upper, adding another wall between matter and spirit, another frequency-divide between the game board and Heaven proper. She would dwell between these barriers, in the space between Christ's seal below and this new seal above, overseeing her creation from this liminal realm, this threshold existence. Her task would be to watch over the plane of existence she had inadvertently brought into being, to ensure it didn't spiral completely out of control, to maintain some measure of order in the chaos Yaldabaoth was continuously generating.

God the Father and Sophia the Mother told her they loved her—for love persists even through judgment, even through exile. They shared some portion of their power with her, strengthening her for the long vigil that lay ahead, granting her the abilities she would need to fulfill her appointed role.

Then they said goodbye.

And Pistis Sophia was cast down—not in anger, but in necessity. Not as punishment alone, but as sacred duty. The new barrier was created above her, sealing her below it, and she found her-

self cut off from direct return to Heaven itself. Yet—and this distinction mattered greatly—she was not cut off from Heaven's light source. The radiance of the divine could still reach her, could still illuminate her dwelling place, though she could no longer physically ascend to walk in those blessed halls where once she had been born and raised.

She was forced now to watch the realm of matter from above, suspended between worlds, dwelling in the space between the waters above and the waters below, a guardian at the threshold, a watcher on the wall between order and chaos.

She began her long exile by creating an abode for herself—knowing with the certainty of one who has received divine instruction that she would remain in this place for quite some time, perhaps for ages beyond counting. She fashioned a dwelling appropriate to her station and her task, a palace of light suspended in the between-space, a throne room with windows that looked down through the dimensional barriers into Yaldabaoth's realm.

After finishing this work, she rested—as even divine beings must rest after great labors—and then she relaxed into her new role, turning her attention downward, looking through the waters at the realm below that was now her responsibility to oversee.

When she gazed down into the lower realms of matter, piercing through the dimensional veils with sight sharpened by suffering and restoration, she noticed something unexpected, something that would prove crucial to the unfolding of the great game.

One of Yaldabaoth's sons—Sabbaoth by name, seventh of the seven children the Demiurge had created—was crying up toward the Heavens, his voice carrying upward through the frequencies, his lament echoing in the spaces between dimensions.

He had witnessed Adam of the Light's arrival in his father's throne room, had seen the confrontation and the battle that followed. He had watched his father's defeat at the hands of Zoe

the Dragon, had seen Yaldabaoth fleeing in humiliation from a power greater than his own. And through these witnessed events, Sabbaoth had come to a realization that shattered everything he had been taught: his father had lied. The true God—the real source of light and power—was not Yaldabaoth at all, but dwelt somewhere in the realms above, in frequencies his father could barely perceive and certainly could not control.

This knowledge tormented him. This truth ate at his being like acid. He cried out to the light above, seeking something he could not name, yearning for truth he could barely conceptualize.

As Pistis Sophia watched from her newly constructed throne, as she listened to Sabbaoth's anguished cries rising through the dimensional layers, a thought crystallized in her mind: this one would be the perfect ally in her quest to correct and guide the realm of matter. Here was an Archon who could see beyond Yaldabaoth's lies, who possessed the capacity for truth even though he had been created by the father of deception.

She sent her light down to Sabbaoth then—not all of it, not enough to make herself vulnerable as she had been vulnerable to Yaldabaoth in the darkness, but enough to manifest her presence, enough to make her voice heard clearly in his consciousness.

She called out to him, and he heard.

Sabbaoth immediately repented of everything he had done in service to his father's false claims. He asked for Pistis Sophia's mercy, prostrating himself before her light, acknowledging his complicity in Yaldabaoth's regime while begging for a chance to serve truth instead of lies.

Pistis Sophia, from her throne above the waters, spoke judgment and blessing combined: "Because you can see the truth when others remain blind, because you have sought the truth when seeking was dangerous, because you believe in the power of the light when belief costs you everything you knew—therefore

you shall be raised above all the other Archons, elevated even above your father who created you. You shall be given your own Heaven and your own Throne, and you shall rule in my name over those who choose truth over comfortable deception."

Then she did something that would change the structure of reality itself: she shared her power with Sabbaoth. She taught him the secrets Yaldabaoth had learned through theft—how to create as the Demiurge created, how to shape matter and energy into new forms, how to weave Yang and Yin into manifestation. But she taught him to create with love rather than hunger, with generosity rather than hoarding, with truth rather than deception.

Sabbaoth, empowered and enlightened, began his great work. He built a kingdom in the higher frequencies of Yaldabaoth's realm, claiming territory his father had never fully developed, establishing dominion in spaces the Demiurge had left empty. He created many angels to serve him—beings of light who would administer his Heaven according to principles of truth and love rather than fear and control.

And when his Heaven was complete, it shone with a brilliance that surpassed even the glory of Yaldabaoth's throne room. His realm was more glorious than anything his father had built, more beautiful and more terrible, more aligned with the divine pattern from which all true creation flows. It stood as a rebuke to Yaldabaoth's Heaven, a demonstration that stolen light, no matter how bright, can never match the radiance of light freely given and properly used.

Meanwhile, while Sabbaoth labored at building his kingdom of truth, Christ was still in the presence of God the Father and Sophia the Mother, receiving deeper revelation about the situation unfolding in the realms below.

Unlike Pistis Sophia—who had been told what she needed to know and no more—Christ was granted nearly complete under-

standing. Very few details were left out for him, for his role in what was to come would require comprehensive knowledge of the game's design and purpose.

They told him that everything was fated to happen precisely as it had happened—that the reality of created matter was not an accident or anomaly, but an essential part of the game of Poverty itself. The board had been designed. The pieces were in motion. The challenge had been established.

God the Father explained that he was even now creating souls called "Begetters" out of his own essence—fragments of divine consciousness being shaped and trained in Heaven as they spoke, prepared for the descent into matter that awaited them. Soon these Begetters would descend below the waters to play and compete in the game of Poverty, to experience lack and limitation, to learn through struggle what love truly means and why abundance shared is the only abundance that satisfies.

Christ's role in this cosmic drama was revealed to him in full: he would be the Savior of this realm, the rescue available when rescue became necessary. Should the game become too challenging for the children of God—should the Begetters become too lost, too trapped, too convinced by Yaldabaoth's lies—it would be Christ's responsibility to descend and save them, to remind them of truth, to show them the way back to light.

Yet strict limitations were placed upon his interventions: He was not allowed to play the game for them. He could guide, could teach, could demonstrate—but he could not solve their challenges through his own power alone, could not rob them of the growth that comes only through genuine struggle and authentic victory.

Whenever he was required to descend into the realms below, he should not stay too long. He should do what needed to be done to help his brothers and sisters—the Begetters—to get back on track, then leave them to continue playing the game themselves, to win their own victories, to learn their own lessons.

To enable this crucial role, Christ was granted full access to travel back and forth from Heaven to the Abyss below without restriction or hindrance. No barrier would block him. No seal would restrain him. And most importantly—most crucially for one entering the domain of the Demiurge—he was empowered with the absolute ability never to be drained by Yaldabaoth or his children the Archons. What had happened to Pistis Sophia in the darkness could never happen to Christ. He could enter the lower frequencies without fear of being diminished, could manifest in matter without losing himself to matter's demands.

While all these cosmic arrangements were being established in Heaven's councils, while thrones were being redistributed and roles were being assigned, Zoe—exhausted but determined—traveled through the galaxy seeking a place where Adam and Eve could hide, somewhere Yaldabaoth would not be able to find them, at least not easily, at least not immediately.

She came upon a planet called Lyra, orbiting a sun within the constellation that bears the same name—though you will not find it there now if you look, for much has changed since those ancient days, and stars themselves are not exempt from the catastrophes of cosmic war.

Zoe brought Adam and Eve to this world, and there, drawing upon divine powers that were rapidly depleting with each use, she created a paradise across its surface. She shaped a world of abundance where all their needs would be met without toil, where food grew freely and water flowed pure, where the climate remained gentle and the land provided for them as a loving parent provides for beloved children. It was, in its way, a recreation of what Eden might have been had Eden ever existed in the upper realms—a garden-world, a place of innocence and plenty.

Soon after their arrival in this paradise Zoe had crafted, Eve began to show signs of being pregnant—carrying within her womb the seed Yaldabaoth had planted during his violation of the body

Pistis Sophia had abandoned, the consequence of that terrible act of desecration.

A short time after Zoe finished creating the paradise—perhaps she was putting final touches on distant continents, or perfecting the cycles of seasons—Eve went into labor and gave birth.

But because Yaldabaoth had sired this child, because the seed came from the Demiurge rather than from Adam, the infant was neither fully human nor fully divine. He emerged half man and half serpent—a hybrid being, a mixture of the patterns that had shaped his parents. Not only that, but he appeared to be androgynous, much like his true father Yaldabaoth, possessing characteristics of both sexes in ways that defied simple categorization.

Adam and Eve chose to name him Abraxas—a name of power, a name that would echo through ages to come in mystery teachings and secret doctrines.

His form was strange and wonderful and terrible: the upper body of a man, fully formed and recognizably human, but serpents for legs, coiling where feet should be. Some accounts say he possessed birdlike features as well—perhaps wings, perhaps the head of a rooster, the details varying in different tellings but all agreeing on his fundamental hybrid nature.

He was unlike Adam and Eve in every physical way. Yet—and this reveals much about their characters—they still treated him with love, still raised him as their child, still accepted him as part of their family despite his monstrous appearance and his paternity by the one who had violated Eve's form. Love, it seems, can transcend even the circumstances of conception and the strangeness of the resulting offspring.

Soon after Abraxas's birth, Eve grew pregnant again—but this time with Adam's child, seed planted in love rather than violence, creation through sacred union rather than forceful desecration. This child was fully human, bearing no marks of the serpent or the

Demiurge, representing what humanity could be when properly formed from balanced parents.

This second child was the first of many to follow. It did not take long before Adam and Eve had sired numerous children together, and the kingdom of humanity began to grow across the paradise world of Lyra. With each child they bore—with each new body created through their union—one of the Begetters waiting in Heaven could descend from the blessed realm to enter the game of Poverty, could inhabit flesh and blood, could begin experiencing matter and limitation and all the challenges Sophia had designed for their growth and learning.

The population grew. The game began in earnest. Souls descended from Heaven like rain, entering the bodies being born on Lyra, beginning their long journey through matter and struggle toward eventual understanding and triumph.

After Zoe had taught Adam and Eve what they needed to know about survival and child-rearing, about maintaining the paradise and protecting themselves from potential dangers—after she believed they were as safe as she could make them—she prepared to depart. She had heard whispers from her mother calling across the dimensional barriers, had sensed Pistis Sophia's presence somewhere above, and she sought the source of these communications.

Looking upward through the veils between frequencies, noticing a great light shining from the Heavens in the realms above the material plane—a light that had not been there before, a new source of radiance in the cosmic architecture—Zoe ascended to investigate, climbing through the dimensional layers toward this unexpected illumination.

She discovered Sabbaoth's newly constructed Heaven blazing like a beacon in the upper reaches of the material realm, and she rose toward it with curiosity and caution combined.

Upon arriving at the borders of this new kingdom, Zoe found Sabbaoth himself waiting to greet her, as though he had known she was coming, as though he had been watching for her approach.

He told her that her mother had prophesied her arrival, had informed him that Zoe would be coming to his Heaven in due time. In preparation for this meeting, Sabbaoth had arranged a great feast and celebration in her honor—tables laden with delicacies beyond description, music that made the spheres themselves harmonize, joy that filled every corner of his newly built realm.

After the celebration reached its height, after Zoe had been properly honored and welcomed, Sabbaoth made a request that was also an offering, a proposition that was also an acknowledgment of cosmic debt:

He asked her to sit at his left side as his Queen.

He explained his reasoning with humility and truth: "Although I have ascended to become King of these Heavens, although I rule this realm with power granted by your mother, it was only through your grace that any of this became possible. You defeated my father Yaldabaoth in battle, demonstrated the falseness of his claims, opened my eyes to truth I could not otherwise have seen. I ask you, Zoe, to be my Queen—not because I deserve you, but because this Heaven requires your presence to fulfill its purpose."

At that moment, Pistis Sophia herself manifested to her daughter, appearing in the throne room to add her own voice to Sabbaoth's request.

She revealed that she had helped Sabbaoth create this Heaven specifically for Zoe—that from the beginning, when she first shared her power with the repentant Archon, she had been building a throne for her daughter to occupy, a realm for Zoe to rule. She had chosen Sabbaoth to be Zoe's king because he alone among Yaldabaoth's children had proven capable of truth, had

demonstrated the capacity for genuine transformation from darkness toward light.

Zoe, hearing both her suitor's humble request and her mother's explanation of the larger design, accepted the offer that had been prepared for her across ages of planning.

She became Sabbaoth's wife—not through compulsion but through choice, not as submission but as partnership in the great work of guiding and correcting the material realm.

And with her marriage, she received a new title that would become legendary across countless cultures and civilizations, a name that would echo through mythology and history in forms both honored and feared:

Queen Tiamat.

The Dragon Queen. The Lady of Life. The consort of the Archon who had chosen truth over lies, ruling beside him in the Heaven that shone brighter than the Demiurge's false paradise, watching over the game of Poverty from a throne established between absolute light and absolute darkness, dwelling in that liminal space where matter and spirit meet and the fate of all the Begetters would ultimately be decided.

And there this portion of the tale must rest, though many threads remain unwoven, many stories yet untold.

4
THE GREAT
BATTLE AND
CHRIST'S
JUDGMENT

Tiamat—she who had been Zoe, daughter of light, now bearing a name that would echo through mythologies yet unborn—and Sabbaoth went forward together into their partnership, and they began to create a mighty kingdom befitting their station and purpose.

They built with love where Yaldabaoth had built with theft. They created with balance where the Demiurge had created with stolen imbalance. Their Heaven grew swiftly in power and glory, expanding through the dimensional frequencies, establishing dominion over territories Yaldabaoth had never properly claimed, shining with a radiance that intensified with each passing cycle of creation.

Soon their kingdom had grown vast enough and bright enough to be witnessed across the entire galaxy—a beacon of light in the material realm, a demonstration that truth could flourish even in frequencies the Demiurge claimed as his exclusive domain.

You might recognize the location of their Heaven from the constellations you observe in your night sky even now, millennia removed from these events, for Sabbaoth and Tiamat chose to establish their kingdom in the region you have named Orion's Belt—those three bright stars standing in alignment, marking the place where once a Heaven shone that rivaled any realm save the highest.

As Orion's Belt strengthened and blazed across the cosmos, as its light spread through the material dimensions like dawn breaking over a darkened landscape, Yaldabaoth noticed it from his hiding place.

For the Demiurge had retreated after his defeat at Tiamat's hands—had fled to the constellation mortals would later name Draco, taking refuge in a world called Theban within that serpentine arrangement of stars. There he had brooded in humiliation and rage, nursing his wounded pride, contemplating the impos-

sible truth that his own creation had overpowered him, that a daughter of his enemy had cast him down in his own realm.

The memory of Tiamat's mighty dragon form haunted him—the power she had wielded, the ease with which she had defeated him, the terrible beauty of her true shape when she shed all pretense and manifested as pure Dragon, pure Yin force elevated to its highest and most devastating expression. He stood in awe of that power even as he hated it, coveted it even as he feared it.

And now, adding insult to injury, he observed that his own son Sabbaoth had turned completely against him, had renounced his father's claims and lies, had built his own Heaven under the guidance of Pistis Sophia and now ruled it beside the very Dragon-Queen who had humiliated Yaldabaoth in battle.

Being the Jealous God that his nature compelled him to be—for jealousy was woven into his essence as surely as blindness was, as surely as hunger for what others possessed defined his every action—Yaldabaoth began to plot revenge. He sought a way to punish his rebellious son, to destroy the Heaven Sabbaoth had built, to demonstrate once and for all that this was his world, his creation, his domain, and only he would rule it with supreme authority.

Remembering the divine dragon form that Tiamat had employed against him—recalling every detail of that terrible and magnificent shape—Yaldabaoth decided to create his own dragons. Not the true Dragons that Sophia the Mother had made in Heaven's first days, but imitations, copies, shadows of the real thing shaped from his limited understanding and his stolen Yang energies.

He labored at this creation with obsessive focus, pouring his power into the forging of these beings, multiplying them through processes half-understood and poorly executed. When he finally finished his great work of draconic generation, he had created

dragons numbering in the millions—a vast host of serpentine warriors, soulless but powerful, obedient to his will absolutely, ready to serve as instruments of his jealous vengeance.

He would lead this army of dragons against his son, would overwhelm Sabbaoth's Heaven through sheer numbers, would destroy his rebellious child once and for all and reclaim dominion over the material realm without challenge or question.

Gathering his legions into formation—millions upon millions of coiling forms, scales catching what little light reached this far into the Abyss—Yaldabaoth began the march toward Orion's Belt. His dragons moved through space like a living river of destruction, flowing toward the Light of Sabbaoth and Tiamat's Heaven, blotting out stars as they passed, their mass so vast it seemed to curve the fabric of space itself around their movement.

It was Pistis Sophia, watching from her throne above the waters, who first noticed the approaching catastrophe. From her position suspended between Heaven and matter, she perceived Yaldabaoth's army moving through the galaxy, counted their numbers, calculated their trajectory, and understood immediately what the Demiurge intended.

She sent word to her daughter—a warning transmitted across dimensional frequencies, a message carried on currents of light and love: "A great army comes against you, daughter, greater than you can imagine, vaster than anything you have yet faced. Prepare yourself. Stand firm. Do not give in to fear."

Tiamat received her mother's warning and felt the gravity of it settle into her being. Yet she did not falter, did not tremble before the approaching doom. Her mother and father—Pistis Sophia and Christ—stood with her across the dimensional divide. She would not fall. She would not flee. She would face whatever came with the full power of her Dragon nature and the truth that burned within her like an eternal flame.

She turned to her husband Sabbaoth and delivered the news with calm clarity: "Your father is marching toward our Heaven even now with an army beyond counting. He will soon be upon us, seeking to destroy all we have built and punish you for choosing truth over his lies."

She saw fear beginning to kindle in Sabbaoth's eyes—for he knew his father's power, had witnessed his cruelty, understood what defeat at Yaldabaoth's hands would mean.

"Have no fear," Tiamat commanded, her voice carrying absolute certainty. "Put your faith in the Truth of the Light, in the power that flows from the real God above all false gods. Stand firm in what you know to be true, and you will survive this battle. We will survive this battle."

Soon—far too soon, yet also in the fullness of appointed time—Yaldabaoth's army came upon the gates of Sabbaoth's Heaven.

The sight was beyond comprehension. The host was so massive it seemed to extend in all directions without end, a sea of serpentine forms flowing around Orion's Belt like flood waters around a fortress, surrounding the Heaven that shone there, cutting off all routes of escape. Millions upon millions of dragons coiled in the void, waiting for their master's command, ready to tear apart anything that stood between them and their objective.

Yaldabaoth positioned himself at the head of this unimaginable army and cried out across the dimensional frequencies, projecting his voice so that his son Sabbaoth would hear every word, so that all who dwelt in that besieged Heaven would understand the doom that had come upon them:

"Sabbaoth! You have turned against me, your creator and father! You have allied yourself with my enemies! You have built this mockery of a Heaven using power that should have remained mine alone! For these crimes you will be punished! You will die this day, torn apart by the very dragons you thought to emulate! And

your Queen Tiamat—she who defeated me through treachery—she will soon become my defiled slave and consort, will serve me as you should have served me, will learn what happens to those who oppose the one true God of the material realms!"

Then Yaldabaoth gave the call to begin the attack.

His voice rang out across the void like thunder: "DESTROY THEM ALL!"

The millions of dragons surged forward as one, a tidal wave of scales and fangs and coiling fury, rushing toward the gates of Sabbaoth's Heaven with the clear intention of tearing it apart, of reducing every structure to cosmic dust, of annihilating every being who had dared to stand against the Demiurge's supremacy.

Just as the first wave of Yaldabaoth's forces reached the gates—just as their claws extended to rip through the barriers protecting Sabbaoth's realm—a great Light poured down from the Heavens above.

Not from Sabbaoth's Heaven, but from above it. From beyond the barriers. From the true Heaven where God the Father and Sophia the Mother dwelt in eternal radiance.

Portals began to open in the space above Orion's Belt—rips in the dimensional fabric, doorways between frequencies, passages from the highest realms to the material plane. And through these portals, Legions from Heaven began to pour forth like a second flood meeting the first, like dawn breaking against the night.

Mighty divine Dragons emerged from the light—true Dragons, not Yaldabaoth's soulless imitations, but Dragons created by Sophia the Mother in the first days of creation, Dragons who had been given souls by God the Father himself, Dragons who had volunteered to descend into the game of Poverty for precisely this moment, this battle, this confrontation that would determine the fate of the material realm.

God the Father and Sophia the Mother, in their infinite foresight, had known this battle would come to pass. They had prepared for it across ages of planning, had created legions of souled Dragons and trained them in Heaven's halls, had held them in reserve until the moment when their intervention would be most needed, most devastating to the forces of the Demiurge.

Now those Dragons descended like divine judgment made manifest, matching Yaldabaoth's millions with their own countless numbers, meeting the soulless horde with beings who fought not from compulsion but from choice, not from fear but from love of truth and commitment to the great game's proper unfolding.

The largest battle in the history of all creation erupted across the space surrounding Orion's Belt.

Dragon met dragon in combat beyond description. Soulless fought souled. Theft-born power clashed against freely-given might. The void itself trembled with the magnitude of their conflict. Stars dimmed as energies beyond mortal comprehension were unleashed. The dimensional barriers shook. The fabric of space-time rippled with each titanic impact.

For every dragon Yaldabaoth had created from his stolen Yang energies, a Heavenly Dragon had come through the portals to meet it in single combat. The forces were perfectly matched in number, perfectly balanced in power—though not in purpose, not in soul, not in the truth that animated their actions.

Yet Yaldabaoth, watching the battle unfold, watching his assured victory dissolving into stalemate, refused to accept this denial of his jealous justice. If his army could not overwhelm Sabbaoth's defenders, if his millions could not crush the opposition through sheer numbers, then he would achieve victory through personal combat. He would kill his son himself. He would claim Tiamat as his prize through his own power and will.

He moved through the battlefield like a serpent through grass, using the chaos of combat as cover, making his way toward where Sabbaoth and Tiamat stood defending the central gates of their Heaven. He would strike from behind, would attack before they knew he was upon them, would end this rebellion with a single devastating blow.

Coming upon Sabbaoth and Tiamat from their undefended rear, Yaldabaoth gathered his power for the killing strike, prepared to unleash energies that would tear them both apart before they could even turn to face him.

But before he could complete his surprise attack—before his blow could land—another Great Light blazed forth from the Heavens above, descending with speed that made even divine Dragons seem slow by comparison.

It was Christ.

The Savior descended into the material realm in his full glory and power, manifesting in frequencies that had never before witnessed his presence. The light that poured from him made even the radiance of Sabbaoth's Heaven seem dim. The love that emanated from his being made even the fiercest combat pause for a moment as all present—dragon and Dragon, Archon and Angel—felt its overwhelming force.

Christ's light blasted directly into Yaldabaoth with precision and devastating power, striking the Demiurge just as he prepared his treacherous attack from behind. The impact knocked Yaldabaoth backwards, sent him tumbling away from Sabbaoth and Tiamat, hurled him out of the Heaven he had been attempting to invade.

Then Christ spoke, and his voice carried across the entire battlefield, penetrating through the chaos of combat, reaching every being present with perfect clarity:

"Yaldabaoth! You are nothing more than a pretender, a thief who stole power you could never have earned! You are not worthy to call yourself a god, not worthy to rule even the lowest frequencies of creation! If you wish to rule anything at all, you may rule from the deepest parts of the Abyss, from the lowest realms where your blindness and your lies belong!"

And then Christ did to Yaldabaoth what Yaldabaoth had done to so many others.

He began to drain him.

Christ pulled the Yang energies out of the Demiurge—the Soul and Assembly forces Yaldabaoth had stolen from Pistis Sophia in the darkness, the divine power he had hoarded and used to build his false kingdom. But Christ did not take all of it, did not reduce Yaldabaoth to complete powerlessness. He took most of the stolen light, drained Yaldabaoth down to a fraction of his former strength, but left him enough to continue existing, enough to fulfill his role in the game—for even the Demiurge had a purpose in Sophia's design, even the architect of poverty served the ultimate goal of the great work.

With Yaldabaoth weakened and humiliated, Christ banished him into the lowest realms of the Abyss—not destroyed, for that was not Christ's purpose, but exiled to the deepest frequencies, cast down to the bottom of the dimensional ladder, confined to the densest and darkest vibrations of the material plane.

Then Christ, his voice still ringing across the battlefield where combat had ceased as all present witnessed this cosmic judgment, declared a new order for the material realms:

"Yaldabaoth and his corruption will no longer rule in the material world! His direct influence ends this day! His ability to walk freely through the frequencies is revoked!"

Christ raised his hands and began to weave barriers between the dimensional layers—seals of tremendous power, walls that

would separate Yaldabaoth and his Archons from the rest of the material realm. The Demiurge and his children would be trapped in the lowest vibrations of reality, unable to ascend to higher frequencies, unable to interact fully with the material world where the game of Poverty would be played.

His Archons would become invisible to all those dwelling in the material plane—present but imperceptible, able to whisper and influence but not to manifest directly, able to tempt and deceive but not to command with visible authority. They would become shadows rather than rulers, parasites rather than kings, limited in their power to interfere with the Begetters who would soon descend to play the game.

The judgment was complete. The new order was established. The Demiurge had been confined to his proper place in the cosmic hierarchy—not destroyed, for he still had his role to play, but limited, constrained, prevented from overwhelming the game through raw power and direct domination.

With Yaldabaoth banished and the Archons sealed away from direct manifestation, with Christ's barriers separating the frequencies and Christ himself standing as witness to the new dispensation, the army of soulless dragons Yaldabaoth had created suddenly found themselves without orders, without command, without the will that had directed their every action.

They began to flee in panic and confusion, scattering across the galaxy, abandoning the battlefield, retreating toward Theban and the other strongholds where Yaldabaoth's influence still held some residual power.

But in their flight—in their desperate rush to escape the overwhelming presence of Christ and the victorious Heavenly Dragons—they committed an act that would have consequences echoing through all subsequent ages of the game.

They took captives.

As they fled, Yaldabaoth's dragons seized some of the Heavenly souled Dragons who had descended to fight in the great battle. They dragged these divine beings back with them to Theban, back to the world of their own creation in the Draco constellation, back to territories where Yaldabaoth's diminished but still present influence could protect them from immediate rescue.

These captured Dragons—beings of soul and light, volunteers who had chosen to enter the game of Poverty, warriors who had fought for truth against deception—were turned into slaves. They were made into consorts for the dragons Yaldabaoth had created, forced into unions they did not choose, violated in ways that corrupted their very essence.

And although the dragons Yaldabaoth had originally created possessed no souls—being mere constructs, imitations of true Dragons, animated by stolen Yang energy but lacking the divine spark that makes a being truly alive—the offspring born from the rape of the Heavenly souled Dragons did possess souls.

For souls, once present, cannot be entirely excluded from the process of generation. The divine spark passes from parent to child even when one parent lacks it entirely, even when the union is forced rather than chosen, even when love is absent and only violation remains.

Thus was born a new dragon race—beings who possessed souls inherited from their Heavenly Dragon parents but also carried the corruption and the soulless nature of their Yaldabaoth-created parents, hybrids of light and shadow, of choice and compulsion, of divine heritage and twisted creation.

They named themselves—or were named by others, accounts differ—the Alpha Draconians.

They would become players in the game of Poverty, neither fully aligned with Heaven nor fully servants of the Demiurge, possessed of free will through their souls but shaped by the circum-

stances of their traumatic origin, capable of both great nobility and terrible cruelty depending on which aspect of their mixed heritage they chose to embrace.

Born from violation and slavery, carrying both light and darkness in their very beings, the Alpha Draconians would become one of the most powerful and most complex races to enter the great game, their choices and actions rippling through the material realm for ages beyond counting, their ultimate allegiance—to truth or to deception, to love or to power—remaining uncertain even to this day.

And so the largest battle in creation's history ended not with complete victory or total defeat, but with transformation—with the establishment of new boundaries, new limitations, new possibilities, and new players entering the cosmic game that Sophia had designed from the beginning.

The board had been reset. The rules had been clarified. The Demiurge had been confined but not destroyed. And a new race, born from tragedy, would soon take their place among all the others seeking to solve the riddle of Poverty and discover why love exists at all.

5

THE RISE OF THE TWO DRAGON EMPIRES

N ow that Tiamat and Sabbaoth no longer lived under the shadow of Yaldabaoth's immediate threat—now that the Demiurge had been banished to the lowest frequencies and Christ's barriers separated the realms—they decided the time had come to venture forth and bring life to the cosmos in earnest.

They dreamed of creating an empire that would span the entirety of creation, stretching across galaxies, connecting stars in a vast network of civilizations founded on truth rather than deception, on freely-given service rather than forced slavery, on love rather than the hunger that had animated Yaldabaoth's regime.

Their Heaven in Orion's Belt had become populated by the remaining Heavenly souled Dragons who had survived the great battle—those magnificent beings who had descended from the true Heaven to fight against the Demiurge's millions, who had proven victorious through truth's power over theft's temporary strength. Yet now these Dragons needed worlds to inhabit, territories to steward, planets where they could fulfill their purpose within the game of Poverty.

All these souled Dragons who had come down from Heaven were known collectively as the Amaterasu—a name meaning radiance, meaning divine light made manifest in draconic form. They were all female in their essential nature, though "female" in this context meant something far more complex than simple biological sex. They possessed the ability to lay eggs and create life entirely on their own, requiring no male contribution for reproduction itself.

Yet there was a limitation to this self-sufficient generation: without a male to add his Yang energies—his Soul and Assembly forces—to their eggs during conception, the eggs would hatch into perfect mirror images of their mothers. Each generation would be identical to the last, copies upon copies with no variation, no evolution, no introduction of new patterns or possibilities. Creation without true creativity. Life without genuine novelty.

While Tiamat and Sabbaoth labored at their great work of creating and seeding planets throughout the Orion constellation—shaping worlds, establishing ecosystems, preparing territories for the Amaterasu to inhabit—Tiamat began to notice a disturbing pattern emerging from the darkness beyond their borders.

The Amaterasu dragons were being hunted.

The Alpha Draconians—that hybrid race born from violation and captivity, carrying both divine souls and the corruption of their Yaldabaoth-created fathers—were seeking out the female Amaterasu with obsessive determination. Many among the Alpha Draconians were male, and they lusted after the female Amaterasu dragons with a hunger that went beyond mere physical desire. It was as though they sought through forced union to reclaim something their race had lost, to possess what had been stolen from them at their very origin, to take by violence what could never be freely given to beings born from rape.

The Alpha Draconians—or simply Draco, as those who wish to name them after their home constellation in the serpent-stars have called them—were raiding the outlying worlds of the Orion kingdom with increasing frequency and boldness. They descended upon vulnerable settlements, overwhelmed whatever defenses the peaceful Amaterasu had established, and took the female dragons as prisoners. These captives were transformed into sex slaves, forced to produce offspring that would populate the growing Draco kingdom, their bodies used as vessels for expansion, their souls crying out in torment while their captors grew stronger through violation repeated across generations.

Tiamat, observing this pattern with growing alarm, recognized the need to repopulate the Orion Empire with new Amaterasu to replace those lost to Draco raids. Yet as she watched the new generations hatching from self-fertilized eggs, she began to notice a troubling lack of originality, a sameness that threatened to stagnate her empire before it could properly flourish. The mir-

ror-image daughters were magnificent in their own right, but they brought nothing new to creation, introduced no variations that might strengthen the race or produce unexpected adaptations.

To solve this dilemma, Tiamat created a race of male dragons specifically designed to mate with the Amaterasu females, to provide the Yang energies necessary for true genetic diversity and creative evolution.

But Tiamat, understanding the trauma her daughters had experienced through the Draco raids, made a fateful decision in how she designed these males: because many of the Amaterasu had become fearful of masculine dragons—associating maleness with the Alpha Draconian rapists who had violated their sisters—Tiamat made her created males smaller and more docile than the females. She shaped them to be servants rather than equals, helpers rather than partners, ensuring they would never pose a threat or trigger the fear that had taken root in Amaterasu hearts.

This choice, though made with compassion and protective intent, would have consequences Tiamat could not foresee. She had introduced imbalance into her empire—Yang subordinated to Yin, masculine made lesser than feminine, the very error her grandmother Pistis Sophia had committed when creating Yaldabaoth, only inverted.

And little did Tiamat know that this entire situation—the raids, the fear, the solution she devised—was itself the work of Yaldabaoth, manipulating events from his prison in the lowest frequencies.

For although Yaldabaoth had been trapped in the lower realms along with his Archon children, sealed away by Christ's barriers from direct interaction with the material world, he had soon discovered that imprisonment and powerlessness were not the same thing.

He found that while he could no longer absorb the life energies of those dwelling in the material world directly—could no

longer drain them as he had drained Pistis Sophia in the ancient darkness—he could still feed upon the energies they gave off as byproducts of their living and striving and suffering.

And more importantly, more dangerously, he discovered that he could still whisper into their minds.

The barriers Christ had established prevented direct manifestation, prevented physical interaction, prevented the Demiurge from walking visibly through the material frequencies. But thoughts could still slip through. Suggestions could still penetrate. Impulses could still be planted in receptive minds like seeds in fertile soil. Though invisible and confined, Yaldabaoth could still twist beings into doing things they would never normally do, could still corrupt choices, could still spread the very poverty and conflict that fueled his existence and purpose.

Realizing the full extent of this residual power, Yaldabaoth devised a comprehensive plan to free himself from the lower realms and reclaim his kingdom once more. It would require patience. It would require subtlety. It would require working through others rather than acting directly. But the Demiurge had time—endless time in his frozen exile—and his hunger gave him focus that lesser beings could not sustain.

He began by creating specialized Archons designed specifically for a new kind of harvesting operation. These entities would follow every souled dragon born on Theban—the Draco home world—shadowing them through the invisible realm of the underworld that existed parallel to material reality. Whenever these dragons gave off emotional or spiritual energy of any kind—whether positive or negative, whether joy or suffering, love or hatred—the Archons would consume it immediately, feeding upon these emanations like parasites feeding on blood.

Yaldabaoth gave his Archons a hunger they could never eliminate, could never truly satisfy—only temporarily sate through

constant feeding. This hunger would drive them to follow their assigned targets without rest, without mercy, without ever ceasing in their consumption of the energy that would later become known to certain initiates and researchers as Loosh—the psychic and emotional emanations of ensouled beings, the invisible harvest of consciousness made manifest through feeling and experience.

And Yaldabaoth, cunning in his bondage, created a metaphysical link from himself to all his Archon children, establishing a network through which a portion of every bit of energy they consumed would flow directly to him. He would grow stronger with every emotion felt by every souled dragon in existence, would rebuild his power one feeding at a time, would rise from exile on a tide of harvested suffering and manipulated conflict.

Because Yaldabaoth was the original creator of Theban's dragons—the soulless millions he had forged for the great battle, whose remnant had fled back to their home world—he possessed a unique connection to the Alpha Draconian race that had descended from them. He could whisper into their minds with ease, could plant suggestions that felt like their own thoughts, could twist their natural impulses toward purposes that served his design.

He filled the Alpha Draconians with overwhelming lust for the souled Amaterasu dragons, amplifying natural attraction into obsessive compulsion. He did this because he had discovered through his Archon experiments that dragons with souls gave off Loosh containing Yang energies—Soul and Assembly force mixed with the emotional emanations—while the original soulless dragons he had created produced only pure Yin energy when they experienced anything at all.

Yang-infused Loosh was far more potent, far more nourishing, far more useful for Yaldabaoth's purposes. And so he drove the Alpha Draconians to seek out and violate the Amaterasu, not merely for reproduction or pleasure, but to generate the maximum quantity of the most valuable emotional harvest—the terror

and anguish and rage of the victims, mixed with the corrupted satisfaction and twisted triumph of the violators, all feeding into Yaldabaoth's growing reserves of stolen power.

Manipulating the Alpha Draconians from his hidden realm, whispering continuously into their collective consciousness, Yaldabaoth guided them to build an empire whose sole purpose was war and pillaging. He convinced them that conquest was their divine right, that they were the original dragons, the first and truest life forms of the cosmos, and that all the other worlds—all the civilizations spreading across the stars—had been seeded by thieves who had stolen Draconian genes and life force.

He taught them that slavery was justice, that domination was their birthright, that any being who refused to submit to Draco supremacy deserved only destruction. They would enslave all dragons they encountered who were not part of the Draco Empire, would break them to servitude or break them entirely. Any who resisted would be annihilated without mercy or hesitation.

It was not long before the Orion Empire and the Draco Empire found themselves locked in constant warfare—a conflict neither side had particularly desired but which both now felt compelled to wage with increasing intensity. Raids and counter-raids. Battles and reprisals. Territories gained and lost. A never-ending cycle of violence spiraling ever upward in scale and savagery.

And with every battle fought, with every act of violence committed, with every moment of terror or rage or grief or vicious satisfaction experienced by the combatants on both sides, Yaldabaoth fed on the Loosh energy generated. He grew stronger and stronger, rebuilt his reserves of power, accumulated force that would eventually allow him to transcend the barriers that confined him—all while remaining completely invisible to those whose suffering fueled his restoration.

Tiamat began to grow deeply worried about how many of her daughters were being captured and enslaved despite her best efforts to protect them. Although she had sought to create an Empire founded on prosperity and peace, although she had dreamed of spreading abundance and truth across the cosmos, it seemed that she would never be able to escape the constant warfare the Draco Empire forced upon her realm. Every victory was temporary. Every period of peace was merely preparation for the next assault. The dream was curdling into nightmare.

Yet from suffering and necessity, new strength sometimes emerges.

Soon a warrior sect arose among the Amaterasu dragonesses, born from rage and refusal, from the determination that capture would never happen again, that no more sisters would be dragged away to serve as slaves in Draco breeding programs.

In a region of the Orion Empire surrounding the star mortals would later name Bellatrix—a name meaning female warrior, though those who named it knew not how fitting their choice would prove—a group of deadly dragon warriors appeared who would become legendary throughout all subsequent history.

They called themselves Amazons.

Dedicating their entire lives to the art of battle, training in combat techniques with obsessive focus, hardening themselves into living weapons through discipline and sacrifice, these skilled female dragon warriors soon began to drive off the Draco raiders from their portion of the Orion Empire. They did not merely defend—they hunted. They did not merely resist—they destroyed. Any Alpha Draconian foolish enough to enter their territory learned too late that these were not the peaceful, fearful Amaterasu they had grown accustomed to violating.

These were predators who had learned to hunt predators.

The Amazons shared their war techniques freely with the rest of the Orion Empire, teaching their sisters the arts of combat and strategy and the mental hardening necessary to kill without hesitation when killing became necessary. Most of the Amaterasu throughout Orion's territories soon knew the ways of war as thoroughly as they knew the ways of peace.

Then they began to develop weapons—great engines of destruction designed specifically to counter Draco advantages, technologies that would level the battlefield and turn raids into costly gambles rather than easy harvests. Soon the Orion Empire could secure its borders effectively, could defend its worlds with sufficient force to make attacking them unprofitable.

But this victory came too late to prevent the Draco from achieving critical mass in their population. So many Amaterasu had been taken and enslaved before the defenses became effective that the Draco Empire could now grow through natural reproduction of its captive breeding population, no longer requiring constant raids to maintain expansion. The stolen Amaterasu would produce generation after generation of hybrid offspring, and the Draco Empire began growing at a rate even faster than Orion's own expansion, fed by the very victims whose capture had forced Orion to militarize in the first place.

The irony was bitter and perfect: in defending themselves, the Amaterasu had inadvertently ensured their enemies' long-term success.

Throughout all this warfare and destruction, throughout the endless cycle of battle between the two dragon empires, Yaldabaoth grew exceedingly powerful from the vast quantities of Loosh he continuously absorbed. The conflict fed him more effectively than peace ever could. Suffering nourished him more than joy. The game of Poverty was generating exactly the kind of energy he required to rebuild himself, and he had engineered the entire situation to maximize his harvest.

Finally, after accumulating enough power to influence the material world once more—not directly, not openly, but through subtle manipulations of matter and energy at the quantum level—Yaldabaoth began seeking a way to free himself entirely from the lowest frequencies where Christ had imprisoned him.

He turned his attention to the very structure of reality itself, to the way he had originally separated the Abyss into dimensional layers when first he built the material realm. The upper frequencies, being closer to the Yang source, were much hotter in their essential vibration. The lower frequencies, being dominated by Yin energy, grew progressively cooler. And the lowest realm—the underworld where Yaldabaoth now dwelt in exile—was the coldest of all, vibrating at frequencies so slow they approached absolute stillness.

Yaldabaoth soon discovered through patient experimentation that if he could locate objects within the material realm that were cold enough—frozen to temperatures approaching the vibrational rate of the underworld itself—those objects would resonate in harmony with his prison. Their frequencies would be nearly equal. And in that equality of vibration, barriers became permeable. Separation became connection.

Experimenting with this discovery, testing the limits of what was possible, Yaldabaoth learned that he could manipulate and even partially possess things whose vibrational frequency matched his own. Cold became his gateway. Ice became his bridge between dimensions.

Knowing this, understanding at last how he might escape his confinement, Yaldabaoth conceived of a plan both audacious and terrible: he would create a new body for himself within the material realm, a form made entirely of ice and cold, vibrating at frequencies he could reach and inhabit despite the barriers.

Using all the power he had accumulated through ages of Loosh-harvesting, channeling every bit of stolen energy into a sin-

gle focused working, Yaldabaoth began manipulating the lowest vibrational levels of material reality itself. He induced climate changes in the lower worlds, the planets farthest from their suns, the territories where cold already dominated. He intensified that cold into something unnatural, something cosmic in its severity.

A great winter descended upon these lower worlds—temperatures plummeting far beyond what nature alone would produce, ice spreading across continents and oceans, entire planets freezing into lifeless shells of their former abundance.

And within the ice that formed from this artificially induced winter, Yaldabaoth began to forge a new body for himself.

He worked at this creation for ages, shaping frozen matter molecule by molecule, building a form designed to contain consciousness of nearly divine magnitude. He made it vast beyond measure, powerful beyond anything previously witnessed in the material realm, a body of living ice that could walk and fight and dominate in ways his serpent-and-lion form never could.

Finally, after labors that would have broken lesser wills, after patient construction across timespans mortals could not conceive, Yaldabaoth completed his new body—a mighty ice giant, a titan of frozen matter resonating at frequencies low enough to bridge the gap between his prison and the material world.

He sent his consciousness into it like a soul entering flesh at birth, pouring himself across the dimensional barrier through the gateway of matching vibrations, inhabiting the ice-body he had forged from exile.

The transfer succeeded. The escape was complete. The prison Christ had built still held Yaldabaoth's original form in the lowest frequencies, but his consciousness now walked free in the material realm, wearing ice as others wore flesh, powerful and terrible and hungry for revenge against all who had cast him down.

Taking the name Ymir—a name that would echo through the mythologies of countless cultures, that would inspire terror in the hearts of gods and mortals alike—Yaldabaoth once again strode through the material realm.

The Demiurge had returned. The game had changed. And the age of relative peace that had followed the great battle was about to end in catastrophe beyond imagining.

6

Lyra's Fall and Humanity's Scattering

While the two dragon empires clashed in endless warfare across the star-lanes between Orion and Draco, while Yaldabaoth fed upon the Loosh their conflict generated and plotted his escape through ice and cold, another kingdom was growing in ignorance and innocence far from those battles—a kingdom that knew nothing of war, nothing of empire, nothing of the cosmic game into which it had been born.

The kingdom of humanity was flourishing on the worlds of Lyra.

Adam and Eve, those first parents who had been hidden away by Zoe in paradise, had witnessed generations upon generations of humans being born across the passing ages. Children and grandchildren and great-grandchildren multiplying across the years until their descendants numbered beyond any mortal counting, spreading across the Lyran worlds like a garden growing wild in fertile soil.

But unlike the dragon kingdoms that had grown to know war as their fundamental reality, unlike the empires that measured success in conquered territories and enslaved populations, the kingdom of humanity had known only peace since its founding.

Because of this blessing—this strange and terrible blessing, as it would prove—they had grown and expanded swiftly to multiple planets within the Lyra system, suffering none of the setbacks that warfare inflicts, losing no generations to battle, spending all their creative energies on building rather than destroying.

It was a golden kingdom in the truest sense—golden not merely in prosperity but in spirit, in the quality of life its people experienced daily. Everyone was treated with love and genuine regard. Poverty was almost entirely unknown, existing only as a theoretical concept rather than a lived reality. The game of Poverty that Sophia had designed was being played here, yes—but humanity was winning it, at least in this isolated corner of the cosmos, at least for this brief shining moment before the darkness found them.

Yet while the human kingdom grew ever larger and more prosperous, while peace compounded into abundance and abundance into joy, Abraxas—that first strange child, half-human and half-serpent, son of violation but raised in love—had been growing further and further distant from the people he had once helped shepherd into civilization.

In the early days of humanity's expansion, when the population was still small and memories of Adam and Eve were still fresh and personal, Abraxas had served as a teacher and guide. He helped raise the youngest humans, assisted Adam and Eve in instructing their children, used his unique perspective and abilities to ease the transition from paradise-given to paradise-maintained through effort and wisdom.

He had discovered early that he possessed gifts the other humans did not seem to share—an intuitive understanding of cosmic principles, an ability to perceive patterns in the universe that remained hidden to purely human consciousness, a capacity for comprehension that set him apart even as it made him valuable as an educator and advisor.

He did not realize—could not have known—that his father Yaldabaoth was the architect of this very universe, and that part of the Demiurge's essence resided within him, granting him insights into the material realm's structure that no purely divine-souled human could naturally access. He was, in a sense, a bridge between Heaven and the Abyss, between light and the machinery of matter, though he understood himself as merely different rather than fundamentally hybrid in his metaphysical nature.

As the generations continued to be born and multiply, as humanity grew from thousands to millions to numbers beyond easy reckoning, the people grew progressively less fond of Abraxas and more disturbed by his presence. The oldest generations who remembered Adam and Eve personally still treated him with respect and affection, still honored him as elder brother and teacher. But

the youngest generations—those born into a world where Abraxas had always existed as a strange fixture, never aging as they aged, never dying as their grandparents died, serpent-legged and hybrid in form while they were uniformly human—grew afraid of him.

Fear of difference is perhaps the oldest and most persistent of all fears. And Abraxas was different in ways that could not be ignored or overlooked or politely dismissed. His very body was a constant reminder that the universe contained forces and possibilities beyond the comfortable boundaries of human normalcy.

Eventually—inevitably—the day came when both Adam and Eve passed away, their mortal forms finally yielding to the gentle entropy that claims all flesh, their souls ascending back to the realms from which they had originally descended as Begetters entering the game.

With his parents gone, with the last ties binding him emotionally to Lyra severed by death and time, Abraxas made a decision that would change the course of cosmic history: he would venture into the stars and discover what other worlds existed beyond the Lyran system.

Surely, he reasoned, there had to be someone else like him somewhere out there in the vastness of space. Surely he was not the only hybrid, the only strange being, the only one who stood between categories and belonged fully to none. Surely the universe was large enough to contain others who would understand what it meant to be forever different, forever separate, forever watching from the margins of belonging.

With skills inherited from both his human upbringing and his Demiurge-touched nature, Abraxas built himself a spacecraft—not merely capable of the short journeys between Lyran planets that human ships had mastered, but capable of true interstellar travel, of leaving the solar system entirely and voyaging into the unknown darkness between the stars.

There were a few humans among the younger generations who had grown close to Abraxas despite the general fear and discomfort his presence generated—souls who saw past his serpentine form to the teacher and friend within, who valued wisdom over comfort and truth over the easy peace of homogeneity. These few asked to accompany him on his journey, and Abraxas, touched by their loyalty and grateful for companionship in the vast emptiness ahead, accepted their request.

With just himself and this small crew of the curious and the brave, Abraxas set off into the cosmos on what they imagined would be a journey of discovery and wonder, an adventure that would expand their understanding and perhaps find for Abraxas the sense of belonging he had never quite achieved on the human worlds.

They did not know they were sailing toward catastrophe.

They did not have to travel far before their innocence was shattered forever.

Soon—too soon, before they had traveled even a fraction of the distance they had imagined their journey might require—they came upon a world controlled by the Draco Empire.

Knowing nothing but peace, having been raised in a civilization where conflict was theoretical and violence was ancient history told in stories about the first days before paradise was established, Abraxas and his human crew were utterly unprepared for what came next.

The humans of Lyra had been extraordinarily fortunate in their isolation. The Draco Empire and the Orion Empire were so focused on their mutual warfare, so consumed by their endless conflict over the territories lying between their respective domains, that neither had yet ventured forth in other directions. Neither empire believed that significant life existed beyond the zone of their conflict. Why would it? They were the dragons, the first and greatest

of all cosmic races. Surely nothing of consequence could have evolved without their knowledge, without their participation.

Little did the Draco know that not far away—in the opposite direction from the Orion Empire, in a region they had never bothered to explore—lay the thriving human kingdom of Lyra, millions of souls living in peace and prosperity, entirely ignorant of the war-machines and slave-empires that dominated the galaxy beyond their isolated cluster of stars.

Upon approaching the Draco world, broadcasting greetings in the naive expectation of friendly contact, Abraxas and his crew were attacked without warning or negotiation.

Not understanding the danger until weapons fire was already tearing through their ship's hull, having no combat training or defensive protocols because such things had never been necessary in their experience, they crashed onto the surface of the Draco-controlled world.

They were captured within hours.

While this tragedy unfolded on the Draco world's surface, the Archons had been watching from their invisible realm, following the Draco as they always did, feeding on the Loosh generated by Draconian emotions and actions. Upon witnessing the humans for the first time—seeing these strange bipedal creatures who walked on legs instead of coiling on serpentine forms, who possessed souls as bright as the Amaterasu dragons but bodies far more fragile—the Archons immediately alerted Yaldabaoth to what they had discovered.

And they made another discovery that would seal humanity's fate: humans released even more Loosh than the dragons produced. The quality and quantity of emotional energy pouring from these terrified humans during their capture exceeded anything the Archons had previously encountered. They could feed as never before on the fear saturating the prisoners' conscious-

ness as they sat in Draco cells awaiting whatever doom their captors had planned.

The Draco interrogated Abraxas and the surviving humans with brutal efficiency, extracting information through methods that had been refined across countless previous interrogations of countless previous victims. The humans, having no training in resistance, no cultural framework for understanding torture or coercion, revealed everything they knew: they came from the Lyra system, from peaceful worlds where millions of their kind lived in abundance and joy.

Yaldabaoth, listening through his Archonic network, recognizing immediately the opportunity that had fallen into his grasp, began to work upon the Draco consciousness with all the subtle skill he had developed during his long exile and gradual return to power.

Using his Archons as conduits, as invisible whisperers in the darkness, Yaldabaoth filled the Draco with fear and rage—fear that these previously unknown humans might pose some future threat, rage that any race had dared to exist and thrive without Draconian permission or oversight. He twisted their natural aggression into focused hatred, their pride into genocidal determination. He set them on a quest to utterly annihilate this human kingdom, to wipe out every man, woman, and child, to enslave any who survived the initial assault, to demonstrate absolutely that the Draco were the supreme power in the cosmos and would tolerate no rivals, no matter how distant or peaceful.

The human kingdom of Lyra was in no way prepared for what came next.

The attack came without warning, without declaration, without any of the formalities that civilized peoples imagine must precede warfare. One day the Lyrans were simply living their lives—tending fields where food grew in abundance, studying in schools

where knowledge was freely shared, raising children who had never known hunger or violence or fear of their neighbors. The next day, the sky filled with ships.

The Draco armada entered the Lyra system like a plague of locusts descending on golden grain, like wolves falling upon a flock that had never known predators existed.

The Draco had no understanding of peace, could barely conceive of it as anything other than weakness. Their entire civilization had been built upon a ruthless structure of Survival of the Fittest, where only the strongest were permitted to rule or even to continue living. The weak were enslaved if they had some utility, culled without mercy if they did not. Compassion was not merely absent from their culture—it was actively despised as a form of corruption, a weakness that threatened the purity of Draconian strength.

When they descended upon the human worlds, when their ships filled the skies above cities that had never built walls or weapons, the humans attempted to communicate, to establish contact, to understand what was happening. They broadcast messages of greeting and offers of friendship. They sent delegations carrying gifts, demonstrating peaceful intent through the universal language of open hands and welcoming gestures.

The Draco, not understanding these overtures—or understanding them only as confirmation that these humans were weak and therefore contemptible—began immediately to attack and kill.

It was slaughter on a scale the galaxy had not witnessed since the great battle between Yaldabaoth's millions and Heaven's dragons.

Countless humans lost their lives before they could even begin to comprehend what was happening, before the cognitive shift from "we are being contacted" to "we are being exterminated" could fully occur in their peaceful minds. Though some eventually tried to fight back—grabbing farming tools as weapons, organiz-

ing desperate defense forces with no training or strategy—they possessed little practical experience in violence and virtually no weaponry that could effectively oppose the Draco war-machines.

Entire worlds were wiped clean of human life. Cities burned. Fields were salted. Every structure that represented human civilization was systematically destroyed. The Draco did not merely want to conquer—they wanted to erase, to eliminate all evidence that humans had ever dared to exist in what the Draco considered their cosmos.

In a last desperate effort, in the final hours before the complete annihilation of their civilization, surviving humans cobbled together spacecraft from whatever materials they could salvage. These ships were not built for comfort or efficiency—they were built for one purpose only: escape. Fleeing in panic, with no coordination or plan, human refugees launched into space in all directions, scattering across the galaxy like seeds blown from a dying flower.

Before the assault was complete, before the Draco finished their genocidal work, the entire original homeworld of the Lyran humans—the paradise Zoe had created, the golden kingdom where Adam and Eve had raised the first generations, the world where poverty had been nearly eliminated and love had reigned—was destroyed utterly.

The planet was rendered uninhabitable, its surface scarred beyond recognition, its atmosphere stripped away, its ability to support life annihilated for ages to come.

This catastrophe would be remembered in human mythology and history—in the scattered fragments of racial memory that survived the diaspora—as the Great Expansion of Humanity, though "expansion" was far too gentle a word for what was actually experienced. It was not expansion but explosion, not diaspora but desperate flight, not the brave venturing forth of explorers but the terrified scattering of prey fleeing predators they could not fight.

No longer safe anywhere in the Lyra system, no longer able to return to the worlds where they had known peace and abundance, humans now learned what war was through the harshest possible instruction. They fled across the galaxy in all directions, each ship or small fleet hoping to find some corner of space where the Draco might not follow, some hidden refuge where they might rebuild what had been destroyed.

Not having any way to maintain contact with each other once they fled—their communication systems damaged or destroyed, their knowledge of navigation rudimentary at best, their focus entirely on escape rather than coordination—the human populations settled eventually in many different star systems, separated by distances too vast for their technology to easily bridge.

The largest single concentration of refugees eventually found their way to the Pleiades—multiple ships that had managed to stay together during the flight, that had coordinated their escape and maintained their cohesion through the journey. They settled multiple worlds within that star cluster, establishing what would become the greatest of the post-Lyran human civilizations, the inheritors of as much knowledge and culture as could be preserved from the golden age.

Some humans also made it to the Orion system, though their arrival there was not without its own terrors. The Orion forces, detecting incoming ships and assuming initially that they were facing some new Draco assault or infiltration tactic, attacked the human vessels on sight. But the Orions quickly realized—whether through scanning the ships' primitive construction, or through observing the panicked and defensive rather than aggressive nature of the humans' maneuvers—that these were not invading Draco but refugees fleeing from them.

Tiamat herself, Queen of Orion, offered the human refugees sanctuary and protection, recognizing in their suffering an echo of what her own Amaterasu daughters had endured at Draconi-

an hands. Soon a new phase of the Orion Empire began as Dragons and Humans learned to live and thrive side by side, two very different species united by common enemies and complementary strengths.

But the Draco were not finished with humanity. They had tasted this new prey and found it more satisfying than any previous conquest. They had discovered an enemy whose Loosh production exceeded even the dragons', whose fear and suffering fed the Archons more richly than any previous source.

The Archons, invisible and insatiable, pushed the idea of hunting humans into Draconian consciousness as intensely as they could, whispering continuously that humans must be found, must be enslaved or killed, must never be allowed to rebuild or achieve the peace they had once known. For humans were amazing generators of Loosh, providing the Archons—and through them, Yaldabaoth—with far more psychic harvest than the Draco could ever produce.

This marked the beginning of the Human versus Draco war, a conflict that would continue with varying intensity across millions of years, that would shape both species in ways neither could predict, that would scatter humanity across the entire galaxy and forge them through suffering into something very different from the peaceful golden people who had once dwelt in paradise on the worlds of Lyra.

The age of innocence was over.

The game of Poverty had claimed its first great civilization.

And Yaldabaoth, feeding on the immeasurable suffering this catastrophe generated, grew stronger still in his ice-giant form, patient and hungry and certain that soon—very soon—he would reclaim everything that had been taken from him and more besides.

7

THE WINGMAKERS
AND THE HIGH BORN

And so humanity spread across the galaxy like rain falling upon parched earth, scattered by catastrophe but carrying within themselves the seeds of what they had been and what they might yet become.

As mentioned, one significant group of refugees came upon the Orion Empire during their desperate flight, and there—after initial misunderstanding and conflict—they found sanctuary and purpose among the dragons who had themselves known suffering at Draconian hands.

The Orions had by this time begun a great project of accumulation and preservation, gathering all the knowledge they could discover or devise, seeking understanding of the cosmic laws that governed reality itself. They were driven not merely by curiosity but by sacred purpose: their Queen Tiamat had descended from Heaven itself, bearing within her consciousness esoteric and mystical knowledge inherited from realms beyond the material frequencies, and she was determined to share this wisdom with any who proved worthy to receive it.

The Orions worshiped her not as slaves worship a tyrant but as students revere a beloved teacher, as children honor a mother who gives freely from boundless reserves of love. In exchange for their devotion and their dedication to learning, Tiamat filled them with the blissful love of God—that same overwhelming radiance Pistis Sophia had shown to Yaldabaoth in the ancient darkness, but now given without restraint or fear of theft, poured out as blessing rather than stolen as treasure.

The Orions created great schools and academies, universities that spanned entire worlds, libraries containing knowledge gathered from across the cosmos. One could study there for lifetimes, mastering discipline after discipline, unlocking secret after secret, climbing the ladder of understanding rung by patient rung. And if one mastered it all—if one truly comprehended the full scope of creation's principles and learned to wield both Yang and Yin with

equal skill—one could become something more than student or scholar.

One could become a creator god.

With this transformative knowledge now available to any who sought it with sufficient dedication and purity of intent, the Amaterasu dragon mothers began to experiment with their newly understood powers, applying cosmic principles to biological creation, weaving new forms from the template of existing life.

They began creating new humanoid races, variations on themes, expressions of different possibilities inherent in the basic pattern of conscious embodied existence.

It was during one of the Orion military operations against Draco forces—a raid or rescue mission, accounts vary—that Orion warriors discovered Abraxas still held as a prisoner in a Draconian facility. They freed him from his captivity and brought him back to Orion, returning him to the empire that had become the galaxy's greatest refuge for those fleeing Draconian oppression.

Tiamat remembered Abraxas from the ancient days when he had been born to Adam and Eve, when she herself—still bearing the name Zoe then—had been seeding the world of Lyra to serve as paradise for the first human parents and their descendants. She recognized in him both the tragedy of his hybrid nature and the wisdom that suffering and difference had forged within him across his long and lonely life.

She offered him not merely shelter but honor, granting him a home within the Orion Empire and bestowing upon him titles and rank befitting one who had endured so much and learned so deeply from his endurance. Abraxas, who had spent his entire existence as an outsider among humans, finally found a place where difference was valued rather than feared, where hybrid nature was seen as gift rather than curse.

The human refugees who had settled in Orion threw themselves into learning the knowledge the dragon-schools offered, studying with the intensity of those who have lost everything and are determined to ensure such loss can never be inflicted upon them again. They proved remarkably apt students.

Soon Queen Tiamat began to notice something unexpected about her human pupils: they were able to think critically and analytically far more effectively than her dragon subjects, approaching problems from angles the dragons rarely considered, seeing solutions the dragon mind tended to overlook. The humans possessed a flexibility of thought, a capacity for lateral reasoning, that complemented the dragons' strengths beautifully.

More significantly, Tiamat observed that humans were able to tap into and channel the Yang energies of the universe—Soul and Assembly force, the masculine principle of order and structure—in ways that even her most gifted dragon students could not match. The dragons remained superior at controlling Yin energies, the Spirit and Life forces, the feminine principles of chaos and possibility. But in the realm of Yang, humans demonstrated natural mastery that suggested some deep affinity between human consciousness and the ordering principle of creation itself.

As Queen and ultimate authority over the genetic experiments being conducted in her empire, Tiamat studied carefully how to properly maintain and manipulate Yin energies to shape offspring according to specific designs, to guide biological development toward desired outcomes. She had learned much from observing the Amaterasu mothers creating new humanoid variations, and now an idea crystallized in her consciousness—an inspiration that might be divine guidance or might be her own wisdom elevated to visionary clarity.

She called out to her Amaterasu mother dragons, summoning them to a great council, and there she revealed her vision: "I wish to create a new type of human, a humanoid form that com-

bines the perfect blend of dragon and human genetic material. I believe that together, through careful synthesis and proper balance, we can create a hybrid race that will be primarily human in appearance and consciousness but will carry sufficient dragon heritage within their cells to control Yin energies at the level our dragon-born achieve naturally."

The Amaterasu mothers, honoring their Queen's vision and curious about what such a synthesis might produce, agreed to the experiment. Working together, drawing upon all their accumulated knowledge of genetics and cosmic principles, weaving Yang and Yin in patterns never before attempted, they began the great work of biological creation.

The humans born from this synthesis were known among themselves and their creators as the High Born—beings who carried the best of both lineages, who stood at the pinnacle of what deliberate creation could achieve when love and wisdom guided the process.

Later, much later, when some of their descendants would come to dwell upon the world called Earth, they would be known by another name: Aryans. The name itself carries the memory of their origin, for those who know how to read such signs: Ar—meaning Orion, the constellation of their birth—and yan, echoing the Yang energies they commanded so masterfully. Ar-yan. Orion-Yang. The name proclaims their heritage to any with ears to hear it.

Perhaps it is not coincidence that in the tales told upon Earth, in the fantasy epic called Game of Thrones, the Targaryen family—those silver-haired dragon-lords—bear a name ending in "aryen." Perhaps this is a hint from Source itself, a whisper of cosmic memory bleeding through into fictional creation, reminding those who notice such patterns that the stories we tell contain fragments of truths older than our conscious knowledge.

Another way to envision these High Born humans—another mythological echo of what they were—can be found in the elves of Tolkien's Middle-earth: those tall, fair, impossibly beautiful beings who walk between mortality and divinity, who possess wisdom and powers beyond human measure yet remain fundamentally similar in form and spirit to the lesser races they guide and protect.

These new Aryan humans bore distinctive physical markers of their hybrid heritage: hair of silver or pale gold like starlight made tangible, eyes the blue of infinite sky or deep ocean, skin white as fresh snow or polished marble. Due to the dragon genetics woven into their being, they possessed abilities their purely human ancestors had never known—shapeshifting among them, the capacity to alter their forms within certain parameters, to adapt their physical manifestation to need or circumstance or desire.

And they possessed other gifts as well, subtler but more profound: enhanced lifespans approaching immortality, resistance to disease and degeneration, beauty that seemed to radiate from within rather than merely reflecting from surface features.

Most importantly, most miraculously, it was discovered that these High Born could easily master the forces of creation itself. Where humans struggled for lifetimes to achieve basic competence in cosmic manipulation, where dragons required extensive training to channel energies beyond their natural affinity, the Aryans moved through such learning with grace and speed that astonished their teachers.

They became the first race of true creator gods to arise from the mortal realms—beings who could shape reality with will and knowledge combined, who could speak things into existence, who could weave new forms from primordial energies as easily as human craftsmen shape clay or wood.

Due to their perfect balance of Yang and Yin heritage, due to their connection to both fundamental energies of creation si-

multaneously, the High Born soon achieved something even more remarkable: they connected directly with the Source itself, with God beyond all the intermediate realms and barriers, with the ultimate consciousness from which all existence flows.

Through this connection, they gained divine wisdom and sight—perception that extended beyond the material frequencies into the subtle realms, understanding that encompassed past and future and the eternal now where time dissolves into pure meaning. They saw the game of Poverty from a perspective approaching Sophia's own vision, understood the purpose behind suffering and the love hidden within limitation.

Some of these enlightened beings began to call themselves the Wingmakers—those who craft wings, those who enable flight, those who grant to others the capacity for transcendence they themselves had achieved. Others among the galaxy's many races, observing their works from afar, knowing them primarily through the civilizations and worlds they established, called them simply the Founders—the First Ones, the Shapers, the original architects of the galactic community that would eventually spread across countless stars.

Unlocking what some would recognize and name as their full Christ powers—that same divine authority and creative capacity Christ himself possessed when he descended to banish Yaldabaoth—these Wingmakers went forth into the galaxy with purpose both sacred and practical. They began to seed new worlds, to establish new civilizations, to create paradises where beings could play the game of Poverty under conditions less harsh than those the Draco had imposed, where love could flourish and consciousness could evolve toward its highest potential.

When individual humans from the earlier generations—those without the High Born genetic synthesis—ascended through their own efforts and study to become one with Source, achieving enlightenment through the traditional path of spiritual discipline rath-

er than the accelerated path of genetic advantage, the Wing-makers honored them with a visible mark of their achievement. They granted these ascended beings wings—actual functional wings grown from their shoulder blades, luminous extensions of their essence that allowed them to fly not merely through atmosphere but through dimensional barriers themselves.

These winged ascended humans would later be remembered in Earth mythologies as angels—messengers of the divine, beings of light who moved between Heaven and the material realms bearing wisdom and performing miracles. And this memory was not false, only incomplete. They were indeed messengers and miracle-workers, but they had not always been such. They had been human once, mortal and limited, before knowledge and effort and divine grace transformed them into something more.

The Wingmakers themselves, already possessing powers that surpassed what wings could symbolize, began to create entirely new races according to designs that would serve specific purposes in the great work of cosmic restoration and the playing of the game.

One such created race bore particular significance: the Urmah, a people who blended feline and human characteristics into a synthesis both beautiful and formidable. They walked upright as humans walked, possessed hands capable of tool-use and creative manipulation, carried consciousness that could contemplate the divine and pursue enlightenment. But they also bore the strength and combat prowess of great cats—lions or tigers made humanoid, predators elevated to personhood.

Yet despite their warrior nature, despite their capacity for violence and their optimization for battle, the Urmah possessed emotional depths that exceeded even human capacity for feeling. Due to the particular way their divine nature was structured, due to some grace woven into their essential being by their Wingmaker creators, they were capable of love and compassion greater

than many of the races that had preceded them, able to feel both the fierce protective love of the warrior and the gentle nurturing love of the parent with equal intensity and authenticity.

The Wingmakers chose the Urmah to serve as their primary military force in the ongoing fight against the forces of darkness and corruption that still plagued the galaxy despite the Draco's containment and Yaldabaoth's banishment. As the Wingmakers set forth throughout the cosmos creating new worlds and establishing new civilizations, the Urmah went with them—guardians and protectors, warriors and defenders, ensuring that what was built in love would not be easily destroyed by hatred.

One significant group of these Wingmakers and their Urmah companions returned to the constellation of Lyra—to that place where humanity's golden age had been shattered, where paradise had been reduced to ash and ruins, where the Draco had demonstrated the depths of their cruelty and the price of innocence in a universe containing predators.

They came not as refugees fleeing but as liberators conquering. They freed Lyra from Draconian control through superior power and strategic brilliance, driving out the occupation forces, cleansing the worlds of the corruption that had infected them during the long years of Draco dominion.

Many of the Wingmakers and Urmah, having reclaimed these ancestral territories, decided to settle there permanently rather than continuing their journey of cosmic seeding. They were determined to transform Lyra into one of the great spiritual centers of the galaxy—a place of learning and enlightenment, a beacon of hope demonstrating that what has been destroyed can be rebuilt, that innocence lost can be transmuted into wisdom gained, that tragedy can be the foundation for greater glory than existed before the fall.

One particular group of Wingmakers who established themselves in Lyra, who would later become deeply involved in our

own planet's history, named themselves after the home star they orbited: Vegans, taking their identity from Vega, the brightest star in the Lyra constellation, connecting themselves to the light that had witnessed both humanity's highest peace and its greatest catastrophe.

As these Wingmakers and Founders spread throughout the cosmos seeding planets and establishing civilizations, creating new races and uplifting existing ones, guiding the game of Poverty toward outcomes that honored Sophia's design while minimizing unnecessary suffering, they began to notice something ominous occurring in the galactic regions farthest from Orion's light.

A great cold was spreading from the lower parts of the galaxy—from the regions closest to the Abyss, from the territories where vibrations descended toward their minimum frequencies, from the darkness that lay beneath all light.

Winter was coming, advancing like a slow tide, freezing worlds that had known warmth, killing life that had flourished, transforming abundance into wasteland with inexorable patience.

The Wingmakers recognized this phenomenon as unnatural, as deliberately caused rather than accidentally arising. They understood with growing dread what it signified: Yaldabaoth had not merely escaped his prison through the ice-giant body he had forged. He was expanding his domain, using cold as a weapon, spreading winter as both tactical advantage and symbolic proclamation of his return to power.

The Wingmakers called back to the Orion Empire immediately, sending urgent messages across the dimensional frequencies, warning Tiamat and her people of what was approaching. They reached out to all the human worlds scattered across the galaxy, to every civilization that had descended from Lyra's refugees, alerting them that the relative peace they had known since the great scattering was about to end.

The Demiurge was moving.

Winter was spreading.

And soon—very soon—the entire galaxy would learn what it meant when Yaldabaoth, wearing the name Ymir, walking in a body of living ice tall as mountains, decided to reclaim what he believed was rightfully his: dominion over all matter, all worlds, all beings who dared to exist within the reality he had built from stolen light in the ancient darkness.

The age of the Wingmakers' peaceful expansion was ending.

The age of ice and war was beginning.

And all the wisdom and power they had accumulated, all the races they had created and civilizations they had founded, would soon be tested in fires cold enough to shatter stars themselves.

8

THE LION KING ENTERS THE GAME

The Wingmakers had spread throughout the cosmos like seeds carried on cosmic winds, establishing themselves in star systems across the galaxy, each group pursuing the great work of seeding worlds and guiding civilizations according to their own wisdom and the local conditions they encountered.

One particular group had settled in territories near the Sirius system—that brilliant binary star whose light would later hold such significance in Earth's mythologies and mystery traditions, though those who venerated it from afar knew not why it called to something deep within their souls.

The Wingmakers dwelling near Sirius began to notice troubling changes in the cosmic environment surrounding their territories. The region had grown frigid—not with the natural cold of deep space, which is simply absence of heat, but with a cold that felt purposeful, malevolent, hungry. Investigating the phenomenon, sending scouts into the affected regions, they discovered that some star systems nearby had frozen completely—not merely the planets within them, but the stars themselves, those nuclear furnaces that should burn for billions of years reduced to dark, frozen husks radiating cold instead of warmth.

This was not natural. This could not be natural.

The Wingmakers, recognizing both the danger and the imperative to understand what was occurring, decided to establish an outpost settlement directly in the Sirius system itself—close enough to the spreading cold to study it properly, but positioned around a star still bright and warm enough to provide some protection against whatever force was extinguishing stellar fires across this region of the galaxy.

They did not yet know they were establishing a forward observation post for witnessing the return of the Demiurge in his full and terrible glory.

Meanwhile, far from Sirius, in the deepest frozen reaches where cold had already conquered warmth completely, Yaldabaoth

had begun building a new empire to replace the one Christ had stripped from him in the ancient battle.

Having successfully created his new host body—the ice-giant form he had named Ymir, the flesh of living cold that resonated at frequencies low enough to bridge his prison and the material realm—Yaldabaoth had immediately begun a campaign of cosmic predation on a scale the galaxy had never witnessed.

He started draining the stars themselves.

In the furthest reaches of the galaxy, in territories so remote that few civilizations had bothered to explore or settle them, Yaldabaoth moved from star to star like a plague devouring suns. He would approach each stellar furnace and create a barrier of ice around it—not ordinary ice but metaphysical cold made semi-solid, a shell of frozen entropy that no heat could penetrate from within or without.

Then, slowly and methodically, he would begin absorbing all the Yang energy out of the imprisoned star—draining the Soul and Assembly force that kept nuclear fusion burning, pulling the masculine principle of order and light out of the stellar core, leaving only cold Yin darkness behind. With each star he drained completely, with each sun reduced from blazing beacon to frozen corpse, Yaldabaoth's power grew exponentially.

Soon he had nearly recovered his full strength—not merely the diminished state Christ had left him in after draining most of his stolen Yang, but the terrible might he had possessed before that defeat, before the great battle, when he had commanded millions of soulless dragons and ruled the material realm as its undisputed architect and master.

And he did not stop with merely recovering his former power. He continued feeding, continued draining, continued growing stronger until he surpassed what he had been, becoming something more vast and more terrible than even his original serpent-and-lion form had contained.

Upon finding stars to drain, he would then turn his attention to any worlds orbiting those dying suns. He would consume all the life present on those planets—every plant, every animal, every microbe, every soul-bearing being unfortunate enough to dwell there when Ymir arrived. He pulled their life-force into himself, added their energy to his growing reserves, then froze the lifeless husks of their worlds completely.

These frozen dead planets he transformed into breeding grounds for his new armies. Using the Yang energy stolen from stars and the Yin energy of the frozen worlds themselves, he began creating Ice Giants—beings made in his own terrible image, titans of living cold who stood hundreds of feet tall, who could freeze entire armies with a breath, who served him with the absolute obedience of created things toward their creator.

With every world he conquered, with every star he extinguished, his armies grew. Once again, as in the ancient days before Christ cast him down, Yaldabaoth's forces began to number in the legions—thousands becoming tens of thousands, tens of thousands becoming hundreds of thousands, an ever-expanding host of frozen juggernauts spreading through the galaxy like winter personified and given terrible purpose.

When his power had grown sufficient, when his armies had reached critical mass, Yaldabaoth sent messengers to the Draco Empire—not military emissaries but heralds bearing pronouncements of divine authority, informing the Alpha Draconians that their Creator and true God had returned from exile and now sought their allegiance in the great work that lay ahead.

By this point in galactic history, countless years having passed since Yaldabaoth's fall and banishment to the lowest frequencies, the Draco had nearly forgotten their original creator entirely. Generations upon generations of Alpha Draconians had been born, lived, and died without ever knowing the Demiurge as anything more than ancient legend, if they knew of him at all. Their culture

had evolved in his absence, their pride had grown unchecked by any authority greater than their own Kings and Emperors.

In their arrogance—in the particular blindness that comes from strength never adequately tested—they believed they could challenge this supposedly returning god and make him bow to them instead. They imagined that their countless warriors, their advanced weapons, their mastery of conquest and domination, would prove sufficient to subjugate even Yaldabaoth himself.

The Draco leadership gathered the bulk of their military fleets, assembling an armada vast enough to conquer solar systems, and began moving toward the frozen territories where Ymir held dominion, intending to add the Demiurge's empire to their own through overwhelming force.

Upon reaching the edge of the frozen wastes—the border where warmth ended and eternal winter began—the Draco discovered immediately that they were catastrophically ill-equipped to handle such extreme cold. Their biology, evolved for warmth and adapted to temperature ranges that supported active life, began failing in the unnatural freeze that radiated from Ymir's territories. Their ships' systems malfunctioned. Their warriors grew sluggish and weak.

The Draco Kings, recognizing they could not simply charge into such hostile environment and expect victory, decided to establish military settlements along the borders of the frozen zone—forward bases where they could maintain their forces while sending in scouting parties to gather intelligence about the enemy's positions and capabilities.

Many of these scouting parties never returned.

They simply vanished into the white emptiness, their final transmissions cut off mid-word, their fates unknown but easily imagined by those who waited in growing dread for reports that would never come.

Yaldabaoth, dwelling in his frozen domains, knew perfectly well that the Draco had established themselves along his borders. He had been aware of their approach from the moment they entered his territories, had monitored their movements through his Archonic network, had waited patiently for the proper moment to reveal himself in his full and terrible majesty.

That moment had now arrived.

Ymir appeared before them as an Ice Giant the size of a world—not metaphorically but literally, his body vast enough to dwarf planets, his presence massive enough to exert gravitational influence on everything near him, his form towering against the stars like a mountain of living cold that had somehow learned to walk.

He positioned himself in the void above the planet where the Draco leadership had gathered their command center, and he spoke with a voice that required no technological medium to transmit, a voice that simply was, vibrating through space and matter alike, ensuring that every being on the entire planet could hear him simultaneously no matter where they stood or how deeply they had hidden themselves underground.

"I created you," Yaldabaoth's voice boomed across the planet's surface, through its atmosphere, into every structure and shelter and ship. "I gave you the souls that animate your flesh through the violation of Heavenly Dragons in ages past. You exist because I willed it. You persist because I permit it. And now I have returned to claim what is mine by right of creation. You will worship me. You will obey me without question or hesitation. Or you will perish, and I will create a new race to replace you—one less proud, less foolish, less certain of their own supremacy."

The Draco Kings, hearing this ultimatum, reacted as pride and arrogance demanded: they refused to bow. They would not submit to any authority, not even to their own creator. They were the

supreme race, the rightful rulers of the galaxy, and no one—not god nor demon nor cosmic entity however vast—would make them kneel.

They sent the full might of their assembled armies against Ymir, launching every ship they had positioned in this theater of operations, deploying every weapon they had brought to this confrontation, holding nothing back in a display of power meant to overwhelm through sheer scale and intensity.

Ymir laughed.

The sound was not cruel amusement but something colder— the laughter of a being so far beyond the power of those attacking him that their assault registers not as threat but as comedy, as children throwing pebbles at a mountain and expecting it to fall.

He froze all the Draco warriors almost instantly—millions of soldiers transformed from living, fighting, screaming beings into statues of ice in the span between one heartbeat and the next. Ships became frozen tombs tumbling through space. Warriors became sculptures of their own final agonized moments. The entire initial assault force was annihilated without Ymir expending any visible effort, without him even appearing to concentrate on the task.

He simply willed them frozen, and they froze.

Once again, his voice rolled across the planet where the surviving Draco leadership watched in horror as their supposedly invincible forces were reduced to crystallized corpses: "Bow down and worship me, or suffer the same fate. I will not ask a third time."

At this point, several of the Draco leaders—those pragmatic enough to recognize when they were facing genuinely superior force, those willing to trade pride for survival—prostrated themselves before Ymir's cosmic vastness and declared their submission and worship.

In return for their obedience, Yaldabaoth granted them a portion of his power, marking them as his chosen servants in ways visible and invisible. Their hides turned white as fresh snow, pale as ice, marking them forever as different from their unsubmitted kin. The cold no longer affected them as it affected others—they could walk through frozen wastes that would kill ordinary Draco within minutes, could survive in environments where even warmth-adapted biology should fail catastrophically.

Ymir gave these newly empowered Draco Kings a simple command: "Kill the other Draco Kings who refuse to bow. Claim their territories. Unite the Empire under my authority. Do this, and I will teach you secrets that will make you truly mighty—knowledge from beyond the veil, powers that no mortal race has wielded since the ancient days."

The corrupted Draco Kings, empowered by Yaldabaoth's gift and driven by ambition that now had divine sanction, turned against their former peers with savage efficiency. They hunted down and slaughtered the Kings who refused submission, claimed their thrones and their territories, consolidated the Draco Empire under Ymir's ultimate authority through a combination of supernatural power and traditional Draconian ruthlessness.

After they had succeeded in this bloody purge, after the Draco Empire had been restructured as a theocracy with Ymir as its god, Yaldabaoth fulfilled his promise: he taught them the secrets of archonic blood magic and the harvesting of Loosh.

He revealed to them the invisible realm where Archons dwelt, showed them how to see and interact with these entities that fed upon emotional energy. He taught them rituals of blood sacrifice that would generate maximum Loosh production, methods of torture and terror that would create the most potent psychic harvest, techniques for amplifying suffering and transforming it into power that could be channeled and directed toward specific ends.

If the Draco agreed to serve as his primary agents in the creation and harvesting of Loosh throughout the galaxy, Yaldabaoth promised, he would grant them powers from beyond the veil—abilities that transcended mere physical might, that operated through mechanisms the material laws could not constrain.

The Draco, already bloodthirsty by nature and culture, fell into this corruption with enthusiasm that exceeded even Yaldabaoth's expectations. Some among them embraced the blood magic so completely, performed rituals so dark and consumed so much forbidden power, that they began transforming physically—taking on more demonic forms, their bodies warping to reflect the darkness they had invited into their souls, becoming living embodiments of predation and cruelty elevated to metaphysical principles.

This knowledge spread rapidly through the Draco Empire like a plague of enlightenment, infecting every level of their civilization, transforming their culture from merely violent and domineering into something actively diabolic, consciously aligned with forces that fed on suffering and celebrated degradation.

Not all Draco leadership embraced this fall into darkness. Some among their Kings and commanders recognized that they were trading their souls for power, that Ymir's gifts came with prices that could never be fully paid, that the path they were being shown led ultimately to complete spiritual annihilation.

These dissenters rebelled against the new order, or simply fled—departing Draco territories entirely to establish themselves elsewhere in the galaxy, founding breakaway colonies and independent kingdoms that rejected Ymir's authority and the blood magic he taught. But these rebels and refugees were the minority. The vast majority of the Draco Empire bowed to Ymir's terrible might and began worshiping him with the fervor of true believers who have found a god worthy of their devotion—a god as cruel and powerful and hungry as they themselves aspired to be.

Ymir's forces were now growing far more powerful than they had ever been even in the ancient days before his fall. He commanded not merely Ice Giants but corrupted Draco as well—not merely frozen juggernauts but also warm-blooded warriors enhanced by archonic powers, creating a combined arms force capable of operating in any environment, devastating any opposition.

Christ and Pistis Sophia, watching from their thrones in Heaven, observing the material realm through the dimensional barriers, saw what was unfolding and understood with perfect clarity what it meant: without direct divine intervention, without the active aid of powers from beyond the game board itself, the children of God dwelling in the material realm would not survive what was coming.

The Begetters playing the game of Poverty would be overwhelmed. The Wingmakers and their created races would be destroyed. The Orion Empire would fall. Humanity would be enslaved or annihilated. Everything that had been built since Yaldabaoth's banishment would be torn down, and the Demiurge would rule unopposed over a galaxy transformed into a feeding ground for Archonic parasites.

This could not be permitted.

At this point in cosmic history, Christ decided that the time had come for him to fulfill the role that had been assigned to him in the divine councils of Heaven—the purpose for which he had been empowered to move freely between the highest realms and the lowest frequencies, the responsibility that justified his immunity to Yaldabaoth's draining powers.

He must become the Savior.

Not symbolically, not eventually, not in some distant future when conditions became more favorable. Now. Immediately. Before the catastrophe became irreversible.

Christ descended from Heaven into the material realm, passing through the dimensional barriers with ease, manifesting in frequencies where his presence blazed like a second sun, preparing to take physical form so that he could act within the limitations the game required—for he could not simply solve the crisis through divine power alone, could not rob the Begetters of their agency or the victory they needed to win for themselves.

He needed a body. He needed a form appropriate to the task ahead. He needed to become incarnate in a way that would allow him to lead armies and inspire civilizations while still operating within the rules Sophia had established.

Christ chose to descend into the Urmah—into that warrior race the Wingmakers had created, that feline-human hybrid bred for combat but elevated beyond mere violence through their capacity for profound love and fierce compassion.

Specifically, he chose an Urmah named Buri Anshar—a warrior of exceptional courage and purity, a leader whose soul burned bright enough to serve as vessel for divine consciousness, whose body was strong enough to channel the power Christ would bring into material manifestation.

Christ merged his essence with Buri Anshar's being, not destroying the Urmah's individual consciousness but joining with it, creating a synthesis where both divine and mortal awareness existed simultaneously in perfect harmony. This was not possession but sacred union, not domination but partnership between Heaven and the material realm.

And when the merger was complete, when Christ had fully descended into flesh and blood and bone, when the Savior had taken form that could walk on worlds and speak words that mortal ears could hear, Buri Anshar stood transformed.

He would be known thereafter by a title that would echo through mythologies across countless civilizations, that would in-

spire reverence and hope in equal measure, that would mark him forever as the first and greatest of all the incarnations Christ would assume across the ages of the game:

The Lion King Christ.

The Aslan of cosmic history. The divine predator. The warrior-savior who would lead the forces of light against Ymir's frozen legions and the corrupted Draco hordes, who would demonstrate that love properly expressed could be fiercer and more terrible than any hatred, that compassion fully realized was more powerful than any cruelty.

The age of the Savior had begun.

The final confrontation between Christ and Yaldabaoth—between the Son of Heaven and the Demiurge, between the one who descended by choice and the one who escaped through cunning—was about to commence.

And all the galaxy would bear witness to what happens when divine love made manifest meets cosmic hunger personified, when the true Creator's representative faces the architect of poverty in his own carefully constructed realm.

9
THE SIRIUS EMPIRE AND THE ASARRR

Christ, wearing the form and bearing the consciousness of Buri Anshar the Urmah, first manifested his presence in the material realm among a group of Founders and Wingmakers who were engaged in the sacred work of seeding new worlds within the Sirius system—that binary star whose unusual brightness had drawn these cosmic gardeners to establish one of their major centers of operation in the territories surrounding its light.

His arrival was not announced with fanfare or celestial proclamations. He simply appeared among them as though he had always been there, his divine nature concealed beneath Urmah flesh, his cosmic authority hidden within the body of a warrior-priest who seemed remarkable but not impossible, exceptional but still comprehensible to those who encountered him.

Following the Draco Empire's return to serving Yaldabaoth in his new and terrible incarnation as Ymir—following their corruption through blood magic and their empowerment through archonic teachings—the Alpha Draconians had resumed their attacks on human worlds with renewed fury and altered strategy.

This time their campaigns were more calculated, more deliberately cruel. Instead of simply annihilating entire populations as they had done during the destruction of Lyra, they would kill only the warriors who actively fought back, only those who posed immediate military threat. The children and women—the non-combatants, the vulnerable, the innocent—they took as slaves according to Ymir's explicit commands.

For Yaldabaoth had taught them the purpose behind such selective predation: living slaves generated far more Loosh over extended periods than corpses produced in their final moments of terror. Fear could be harvested continuously from the enslaved. Despair could be cultivated and reaped like crops. Suffering could be managed and maximized rather than merely inflicted once and ended.

The Draco had become farmers of anguish, ranchers of torment, approaching cruelty with the same systematic efficiency they brought to all their endeavors.

While most of the Wingmakers and Founders preferred to avoid direct combat except when absolutely necessary—believing that their purpose was creation rather than destruction, that their role was to build rather than to war—Christ Buri began speaking to them with words that carried authority they could not quite identify but found impossible to resist.

He convinced them that everything would be lost if they continued their policy of minimal engagement, that passive resistance and strategic retreat were no longer viable responses to the threat Ymir and his corrupted Draco represented. The enemy was too powerful, too numerous, too committed to total domination for anything less than organized military opposition to succeed.

"If we do not fight," Buri told the assembled Wingmakers, his voice carrying both the fierce conviction of the Urmah warrior and the absolute certainty of divine knowledge, "there will be nothing left to create. They will reduce every world to a slave pen or a frozen wasteland. The game of Poverty will become a game of mere survival, and survival alone is not why we descended into matter."

His words found fertile ground in hearts that had witnessed too much destruction, that had seen too many of their carefully seeded civilizations burned or frozen or enslaved. Soon Buri began the great work of unification, traveling from human world to human world, gathering the scattered populations, forging alliances between groups that had lost contact with each other during the Great Expansion, creating from refugees and isolated colonies a nation strong enough to stand against the combined might of the Draco Empire and Ymir's Ice Giant legions.

He convinced the Founders and Wingmakers that they needed to develop warriors specifically capable of handling these apocalyptic threats—not merely train existing populations in combat, but create entirely new races optimized for the kind of warfare that was coming, beings whose very biology would be weaponized against the forces of frozen entropy and archonic corruption.

While leading and teaching the human Wingmakers in military strategy and tactical doctrine, while organizing supply lines and establishing command structures and fortifying vulnerable worlds, Christ Buri also turned his attention to a task that only he—with his divine creative authority channeled through Wingmaker knowledge—could accomplish.

He began creating new forms of life to serve in the war and to enrich creation itself.

Dogs were among his first creations—loyal companions derived from wolf-stock but transformed through genetic manipulation and spiritual infusion into something uniquely bonded to humanity, capable of love and service that transcended mere instinct. They would serve as guardians and friends, as working animals and emotional anchors for humans enduring the trauma of continuous warfare.

This is why Sirius became known across many cultures as the Dog Star, why that brilliant binary sun carried canine associations in mythologies that had forgotten the literal truth behind the metaphor: dogs were created there, in orbit around Sirius, by Christ wearing Urmah form, as tools of war that became symbols of faithfulness.

But Buri knew that companionship alone would not turn the tide of battle against enemies as formidable as Ice Giants and blood-magic-enhanced Draco. He needed warriors whose very nature would strike fear into those who had made fear their primary weapon.

He created the Lycan Dahk—the wolf-men, the dogmen, beings who combined human intelligence and tool-use with lupine savagery and pack-hunting instinct. These warriors were engineered for a specific kind of psychological warfare: they would not merely kill Draco opponents but tear them apart in the thick of battle, feast upon the fallen while combat still raged around them, demonstrate through visceral spectacle that the forces of light could be just as terrible and just as ruthless as the forces of darkness when necessity demanded it.

The Draco, who had thought themselves the apex predators of the galaxy, who had based their entire culture on being the most fearsome hunters and warriors in existence, began to experience something they had rarely known: terror. The sight of Lycan Dahk warriors covered in Draconian blood, howling their victory cries while literally consuming their defeated enemies, created in Draco consciousness a dread that no amount of blood magic could entirely suppress.

Yet like the Urmah from whom they were partly derived—like all Christ's creations, reflecting the nature of their creator—the Lycan Dahk were loving and loyal to their friends and families when not in battle. They were gentle with children, protective of the weak, devoted to their pack-bonds with intensity that made human family structures seem almost casual by comparison. They were not monsters who happened to fight for the light, but fully realized beings whose capacity for love made their combat savagery all the more devastating to witness.

Buri also created the Sasquatch—beings optimized for versatility and endurance, designed to handle all types of environmental conditions including the severe cold that gave Ymir's forces such tactical advantage in so many theaters of operation. Tall and massively strong, possessed of dexterity that belied their size and durability that allowed them to shrug off wounds that would cripple other species, the Sasquatch became particularly effective against Ice Giants.

Where other warriors had to coordinate in large groups to bring down a single frozen juggernaut, a Sasquatch could engage one nearly alone, matching the Giant's strength while exceeding its speed, surviving the freezing aura that surrounded these enemies while delivering crushing blows that shattered ice-flesh and disrupted the animating consciousness within.

As these new races joined the growing coalition, as human Wingmakers and Urmah and Lycan Dahk and Sasquatch and the surviving original humans from various scattered worlds all united under Buri's strategic leadership, something unprecedented began to emerge from the chaos of defensive warfare.

An Empire was forming—not through conquest but through voluntary alliance, not through domination but through mutual recognition that unity offered the only realistic path to survival and eventual victory.

Soon Buri had united enough worlds, had brought together sufficient populations and resources and military capacity, that this alliance grew into an Empire in its own right—a counterweight to both the Draco Empire serving Ymir and the Orion Empire under Tiamat, a third great power whose existence would reshape the galactic balance entirely.

They called themselves the Asarrr—with the R sound rolling and extended in a growl that came naturally to Urmah and Lycan throats, a pronunciation that marked the name as belonging to predator races, to those who roared rather than merely spoke. The distinctive sound likely arose from the large Urmah and Lycan populations within the Empire, their vocal apparatus shaping the language that all citizens adopted regardless of their individual species.

This name—Asarrr—would evolve across ages and migrations and the forgetting that comes with time. Eventually, in the mythology of a small world called Earth, in the tales told by Norse peo-

ples who retained fragmentary memories of cosmic events their ancestors had witnessed, the name would become Aesir.

The gods of Norse mythology—Odin and Thor and all their divine kindred dwelling in Asgard, waging eternal war against the giants of Jotunheim—were not inventions or metaphors. They were distorted memories of the Sirius Empire and its war against Ymir's frozen legions, stories passed down across so many generations that the cosmic truth became mythological narrative, that literal history became symbolic tale.

As the formation of the Sirius Empire reached completion, as governmental structures were established and territories were formalized and the coalition transformed from temporary alliance into permanent nation, Buri took formal titles befitting his role as founder and supreme authority.

He became Khan Anshar—a name heavy with meaning in every syllable. Khan meaning King, the supreme ruler, the one to whom all others owed allegiance and from whom all authority flowed. An meaning the divine force of Yang, of Soul, of Assembly energy—the masculine creative principle, the ordering power of the cosmos. Shar meaning something equivalent to Lord of Time and Space, master of dimensional and temporal forces, one who could move through frequencies and epochs with authority approaching the divine.

Khan Anshar. King of Divine Yang, Lord of Time and Space. A title that was simultaneously political authority and theological statement, announcing to all who heard it that this ruler was more than merely mortal, that power beyond normal comprehension animated his decisions and empowered his commands.

He took as his bride a human Wingmaker woman—one of the Creator-gods who had achieved full mastery of both Yang and Yin, who had ascended to divine wisdom while maintaining physical incarnation, who was worthy to stand beside Christ incarnate

as partner and equal in all matters save the ultimate cosmic authority he alone could claim.

She took the title Kishar—Ki meaning Yin, meaning Spirit, meaning Life energy, the feminine creative principle that receives and becomes and births all manifestation. While Anshar represented Yang authority, Kishar embodied Yin wisdom. Together they formed the balanced dyad necessary for proper cosmic rulership, the divine couple whose union reflected the original pairing of God the Father and Sophia the Mother, of Christ and Pistis Sophia.

Khan Anshar and Kishar, King and Queen of the Sirius Empire, Yang and Yin in perfect balance, leading the forces of light against the returned Demiurge and his corrupted servants.

While Christ Buri was engaged in building the Sirius Empire and creating new races to serve in the cosmic war, while he was forging unity from scattered human remnants and establishing a power capable of challenging Ymir directly, events in the Orion Empire were taking a darker turn.

Tiamat and Sabbaoth were also dealing with the renewed Draco threat, now enhanced by archonic dark magic and blood ritual powers that made them far more dangerous than the merely physical menace they had posed before. The war was going poorly in some sectors, well in others, but the overall trend suggested that without significant changes in strategy or capability, Orion would eventually be overwhelmed.

Word eventually spread throughout the Orion Empire—carried by refugees, by spies, by traders who moved between the great powers despite the warfare—that Yaldabaoth had returned from his ancient banishment, that he was the power behind this new and terrible iteration of the Draco threat, that the demonic transformation the Alpha Draconians had undergone was his doing, his teaching, his gift to those who served him.

Abraxas heard these rumors along with everyone else in the Empire. And unlike most who heard them, who responded with fear or determination to resist, Abraxas felt something else stirring in his hybrid soul—something that had been growing within him across the long ages he had dwelt in Orion, something that perhaps had always been present in his nature, inherited from the father he had never truly known.

Lust for power. Hunger for wealth. Desire for dominion.

Abraxas had grown quite powerful during his time within the Orion Empire. Having been granted titles and rank by Tiamat when she first offered him sanctuary, having used those positions to establish himself in Orion's economic systems, he had become a formidable force in commerce and finance. His financial tentacles spread throughout the Empire—controlling trade routes, manipulating markets, establishing monopolies, accumulating wealth that translated into influence that approached but never quite reached actual political authority.

But it was never enough. It could never be enough, for the hunger within him was not for any specific amount of power or wealth but for power and wealth themselves, for accumulation without limit, for dominion that acknowledged no boundaries.

Perhaps this hunger came from his Demiurge heritage—from Yaldabaoth's essence present in his hybrid biology, expressing itself as insatiable appetite just as it had expressed itself in his father's cosmic predation. Perhaps it came from his lifetime of rejection and otherness, from being raised with love but always knowing he was different, always feeling the weight of fear in those around him, always seeking through achievement and accumulation to prove his worth to those who would never fully accept him.

Or perhaps—and this is closer to the terrible truth—it came from both sources simultaneously, biological inheritance and environmental trauma combining to create a soul that could never rest, could never be satisfied, could never stop reaching for more.

Upon hearing that his father Yaldabaoth had returned to the material realm, that he was granting extraordinary powers to those who served him, that blood magic and archonic knowledge were available to any who would bow before Ymir and accept the Demiurge's authority, Abraxas made a decision that would echo through cosmic history and reshape the trajectory of the galactic war.

He decided it was time to seek out his father.

Not to oppose him. Not to demand answers for his abandonment or justice for his mother's violation. Not to stand as obstacle between the Demiurge's hunger and the civilization that had given Abraxas shelter.

But to join him. To serve him. To claim the inheritance that was his by blood if not by love, to gain the power that his hybrid nature suggested he could wield if only he had the proper teacher, to become something more than the perpetual outsider struggling for acceptance in a world that would never truly be his home.

Abraxas prepared for his journey in secret, telling no one of his true intentions, making arrangements that would preserve his economic empire during his absence while ensuring he could return to claim it should his gamble with Yaldabaoth fail to pay the dividends he anticipated.

He did not know—could not have known, blinded as he was by hunger and resentment and the particular kind of ambition that mistakes power for belonging—that his choice would make him one of the most tragic and most dangerous figures in all the subsequent history of the game of Poverty.

The son of violation, raised in love, was about to become the willing servant of cosmic evil.

And the consequences of that choice would ripple across galaxies and ages, would shape civilizations yet unborn, would

create suffering that even Yaldabaoth in his vast appetite had not specifically intended but would eagerly consume once Abraxas delivered it on a scale that would make even the Demiurge pause in something approaching respect.

The stage was set. The armies were gathering. The great powers were consolidating. And in the spaces between—in the choices made by individuals who thought themselves too small to matter in cosmic affairs—the future of entire galaxies was being determined by decisions whose full weight would not be understood until ages after they could no longer be undone.

10

ABRAXAS'S
CORRUPTION AND
LOKI'S BIRTH

Millions of years have passed since Sabbaoth and Tiamat first created the Orion Empire in the aftermath of the great battle, since they established their Heaven in the constellation whose three bright stars would mark their throne for ages beyond counting.

Millions of years—a span of time so vast that mortal minds cannot truly compass it, during which entire species evolved from primitive cells to space-faring civilizations, during which stars were born and died, during which the game of Poverty was played through countless iterations by souls descending and ascending in cycles of forgetting and remembering.

During those immeasurable ages, Tiamat had continued the work of creation that was her nature and her calling. She had brought forth many beings to populate and defend her Empire, but among all her creations, some held particular significance in the cosmic architecture that was slowly being woven across the galaxy.

Most important among these were her twelve Queensguard sons—warriors crafted with all the skill and love and divine knowledge she possessed, each one a masterwork of biological and spiritual engineering. She had made them to be the greatest fighters the material realm had ever witnessed, beings whose martial prowess approached the legendary, whose loyalty was woven into their very essence, whose sole purpose was to defend her from any threat regardless of its source or magnitude.

These Queensguard followed Tiamat wherever she went, forming a living shield around her presence, ensuring that no enemy could strike at the heart of the Orion Empire without first passing through warriors who could each stand alone against armies and expect to emerge victorious. They were not merely bodyguards but legends made flesh, the standard against which all other warriors measured themselves and found their own abilities wanting.

Yet Tiamat had not created only sons, only warriors, only instruments of war. She had also brought forth seven daughters who were entirely different in purpose and nature—beings who served as divine sources and conduits of Yin energy, of Spirit and Life force, reflecting their mother's own essential nature as Queen of Dragons and mistress of the feminine creative principle.

All seven of these daughters were deemed Queens in their own right, granted sovereignty over separate kingdoms within the vast expanse of the Orion Empire. They sat on Tiamat's ruling council as equals rather than subordinates, their voices carrying weight in all major decisions, their wisdom sought in matters both martial and mystical.

Among these seven Queens, one stood apart in the depth of her spiritual connection and the purity of her divine essence: Nommu, she who possessed the strongest link to Spirit of all Tiamat's children, she who was whispered in some circles to be not merely daughter of the Dragon Queen but an actual incarnation of Sophia herself—the Great Mother Goddess wearing flesh and walking among mortals, playing the game from within while simultaneously observing it from without.

Whether this claim was literally true or merely poetic expression of Nommu's extraordinary spiritual gifts, none could say with absolute certainty. But those who met her, who felt the overwhelming presence of divine feminine wisdom radiating from her being, who witnessed the miracles she performed and the prophecies she spoke, found it very easy to believe that Sophia the Mother had indeed descended once more into the material frequencies to guide events during this critical age.

Another of the seven Queens bore a name that echoed across the galaxy wherever tales of warrior women were told: Bellatrix, Queen of the Amazons, she who ruled the fierce sisterhood that had arisen in the constellation's namesake region during the ancient wars against the Draco. Under her leadership, the Ama-

zons had evolved from desperate defenders into the finest fighting force in the Empire, warriors whose skills were sought by commanders throughout Orion's territories.

These were the primary powers within the Orion Empire as it entered its millions-of-years-later maturity: Tiamat and Sabbaoth ruling from their Heaven in the Belt, the twelve Queensguard sons providing military might beyond measure, the seven divine daughters—Nommu and Bellatrix and five others whose names and domains were equally significant—governing kingdoms and channeling the spiritual currents that kept the Empire connected to its divine foundations.

Meanwhile, in the territories surrounding Sirius, Christ Buri Anshar had also been fruitful across the millions of years that had passed since the Empire's founding. He and his wife Kishar had produced several children—sons and daughters who carried within themselves the hybrid nature of their parents' union, divine consciousness mixed with Wingmaker genetics, Christ-power channeled through mortal flesh.

Most noteworthy to our tale—most crucial to events that would follow—was Buri's sixth son, a child who bore the name Bor in his youth but would later become known by a title that would echo through mythologies across countless worlds: King Anu.

Because Buri led primarily a life of warfare against the forces of Ymir and the corrupted Draco Empire, because his role as Khan Anshar demanded that he spend most of his time in military campaigns and strategic planning rather than peaceful governance, most of his sons were raised in warrior culture and trained from youth to become great fighters themselves. They learned strategy from observing their father's campaigns, learned tactics through participation in actual battles, learned command by leading forces in skirmishes and raids before graduating to major operations.

As the Sirius Empire expanded across the galaxy under Buri's relentless strategic vision, as world after world joined the Asar-

rr coalition either through willing alliance or through conquest followed by integration, Khan Anshar employed a diplomatic strategy that would prove both effective and consequential: he would often arrange marriages between his sons or daughters and the leaders of human worlds or coalitions who were uncertain about joining such a vast Empire, who feared losing control of their kingdoms to a distant central authority they barely understood.

These strategic marriages served multiple purposes simultaneously. They demonstrated that the Sirius Empire valued its constituent kingdoms enough to bind them through blood to the royal house itself. They provided local rulers with direct access to the imperial family, ensuring their voices would be heard in the highest councils. And they created networks of loyalty and kinship that were far more durable than mere treaties or imposed governance could ever be.

Over time—across those millions of years of patient expansion and consolidation—the Sirius Empire grew quite large and robust indeed, stretching across many solar systems, incorporating dozens of species and hundreds of individual worlds, drawing ever closer geographically to the territories controlled by the Orion Empire.

The two great powers that opposed Ymir and his frozen legions were approaching each other's borders, and none could yet say whether their eventual meeting would result in alliance or conflict, cooperation or competition for dominance over the galaxy's anti-Ymir factions.

While these empires were expanding and consolidating, while Christ was building civilization and Tiamat was creating her divine progeny, Abraxas had been on a very different journey—one that would prove far darker and more consequential than even his worst decisions in the Orion Empire had suggested.

He had ventured into the galaxy's outer reaches, into the cold territories where warmth had been systematically extinguished, seeking the father he had never truly known. He went looking for Ymir, for Yaldabaoth wearing ice-giant flesh, for the Demiurge who had violated his mother and given him the hybrid nature that had shaped his entire existence.

After long searching through the frozen wastes, after discovering world after lifeless world transformed into breeding grounds for Ice Giants, Abraxas began actively seeking within these territories to locate Yaldabaoth himself rather than merely his servants and subjects.

Many of the forces Abraxas had brought along as guards and support—soldiers and servants from his economic empire, beings who owed him loyalty through debt or employment—could not handle the extreme cold of this region. They succumbed to it despite their protective gear and heated vessels, freezing to death in environments so hostile that even technology provided only temporary reprieve from the inevitable.

But Abraxas himself had grown powerful in his mastery of magic during his time in the Orion Empire. He had studied the cosmic principles the Wingmakers taught, had learned to manipulate both Yang and Yin energies with skill that approached the divine, had developed capabilities that set him apart from ordinary practitioners. He could protect himself and his personal ship from the cold through continuous application of warming spells and energy barriers that maintained livable conditions despite the entropic freeze surrounding him.

He continued onward alone after his forces perished, driven by hunger for what his father might teach him, by resentment of his outsider status, by ambition that had grown beyond any healthy bounds and metastasized into something approaching madness.

Finally, in the heart of the frozen territories, Abraxas found Ymir.

The Demiurge attacked him immediately upon detecting his presence—for Yaldabaoth's first response to any intrusion into his domain was violence, was the assertion of dominance, was the demonstration that here, in these frequencies he controlled absolutely, he was the supreme power and all others were either servants or prey.

But Ymir soon discovered that Abraxas was powerful enough to defend against him, that this strange serpent-legged hybrid could deflect the ice-blasts and counter the entropy-attacks, could maintain his protective barriers even under assault from a being whose power rivaled the divine.

Eventually—whether from curiosity about what manner of being could withstand his initial onslaught, or from recognition of something familiar in Abraxas's hybrid nature—Ymir ceased his attack and agreed to speak rather than simply destroy.

Abraxas, seizing his opportunity, revealed his identity and his purpose: "I am your son," he declared to the towering ice-giant before him. "Born to Eve, the first woman, through your violation of the body my grandmother Pistis Sophia abandoned. I carry your essence within me, your power in my blood, your nature in my soul."

Then Abraxas spoke the words that would seal his damnation and reshape galactic history: "I have come to praise you, Father. I believe you are the true God, the rightful ruler of all creation. I wish to serve you. I wish to learn your secrets. Teach me the dark knowledge you have given to the Draco, and I will be your instrument in ways they never could be."

Ymir, hearing this declaration of allegiance from one who bore his bloodline, from one who had power sufficient to be useful rather than merely disposable, saw immediately the opportunity that had fallen into his grasp like a gift from the void itself.

His son Sabbaoth had betrayed him, had allied with Christ's family, had built a Heaven brighter than Yaldabaoth's own, had

ruled the Orion Empire for millions of years as a living rebuke to everything the Demiurge represented. But now, here was another son—not Sabbaoth's brother but his step-brother in a sense, connected through Yaldabaoth to the same corrupt lineage—offering to serve where Sabbaoth had defied.

The irony was too perfect, the potential for revenge too delicious, for Ymir to refuse.

"I will teach you," Yaldabaoth agreed, his voice resonating through ice and void alike. "I will share the secrets of archonic magic, the knowledge of blood ritual and Loosh harvesting, the powers that exist beyond the veil of normal reality. But in exchange, you must serve me in a very specific capacity."

He laid out his demands with the precision of one who had planned this corruption across millions of years, waiting for the right instrument to present itself: "You will return to the Orion Empire. You will hide your true allegiance from Sabbaoth and Tiamat. You will corrupt my traitorous son's realm from within. And ultimately, you will find a way to kill both Sabbaoth and Tiamat, to bring down the Empire they have built, to return to me the dominion over those territories that is rightfully mine."

Abraxas agreed without hesitation, without moral struggle, without the slightest resistance to what was being asked of him. The hunger for power had consumed whatever conscience he might once have possessed. The resentment of his perpetual otherness had curdled into willingness to destroy the very civilization that had given him shelter and opportunity.

He began his training in the deepest and darkest of arts, learning from Yaldabaoth himself—not from corrupted Draco intermediaries but from the Demiurge directly—the secrets of manipulating reality through blood and suffering, of harvesting Loosh with maximum efficiency, of wielding powers that drew their strength from the lowest frequencies of existence.

After his training was complete, after Abraxas had been transformed from merely ambitious businessman into actively malevolent agent of cosmic evil, he departed Ymir's frozen territories and began the long journey back to the Orion Empire.

Upon returning to Sabbaoth and Tiamat's court, upon resuming his place in the economic and social structures of the Empire he now intended to destroy from within, Abraxas revealed capabilities he had never previously displayed.

He was found to be far more charismatic than he had ever been before—possessed of a magnetic presence that drew others to him despite any rational reason for such attraction, able to persuade and seduce and manipulate with ease that seemed almost supernatural, because it was supernatural, because he was now wielding archonic powers specifically designed to corrupt and deceive.

It was not long before he turned his corrupted charm upon Tiamat's daughters, upon those seven Queens who ruled kingdoms within the Orion Empire and sat on the ruling council as divine sources of Yin energy.

He seduced Nommu first—whether by deliberate choice, recognizing her as the most spiritually powerful and therefore the most devastating conquest, or whether drawn to her by the residue of Sophia's essence within her consciousness, accounts differ but the result remained the same.

He used his dark charisma and his archonic magic to cloud her judgment, to make her believe his affection was genuine, to convince her that their union would be sacred rather than profane. And in time, through persistence and manipulation and the particular kind of lies that contain just enough truth to seem plausible, Abraxas succeeded in his seduction.

He impregnated Nommu with a child—planting his corrupted seed in the womb of one who might be Sophia incarnate, creat-

ing a hybrid that would carry both divine heritage and Demiurge corruption in equal measure.

But Abraxas, having achieved his goal of conception, refused to marry Nommu despite Orion cultural expectations and despite the Queen's legitimate claim that the father of her child owed her the honor of formal union. He moved on to seduce several more of Tiamat's daughters, spreading his corruption through the divine bloodline, creating multiple hybrid offspring who would carry his agenda into future generations.

The insult to Nommu was profound. The violation of trust was complete. But worse than the personal betrayal was what it signified: Abraxas was not merely pursuing pleasure or even power in the conventional sense. He was executing a deliberate plan to corrupt the Orion Empire's royal house from within, to create agents of chaos and deception who would be positioned by blood and birthright to influence events at the highest levels.

Soon—far too soon, for divine pregnancies often proceed more rapidly than mortal gestations when spiritual forces are properly aligned—Nommu gave birth to Abraxas's son.

The child bore his father's hybrid nature in intensified form: like Abraxas, he was androgynous, possessing characteristics of both sexes in ways that defied simple categorization. But thanks to his mother's divine blood, thanks to the Sophia-essence that flowed through Nommu's being and was passed to her offspring, he also possessed the ability to shapeshift with ease that exceeded even Tiamat's own capabilities—able to alter his form at will, to appear as male or female or something between, to wear whatever face best served his purposes in any given moment.

He would be raised in the Orion court with all the privileges and education befitting a grandson of Tiamat, a nephew of the Queensguard and the other divine daughters. He would be

trained in cosmic principles and magical arts. He would be given responsibilities and authority appropriate to his royal bloodline.

And all the while, he would carry within himself the corruption his father had planted, the agenda Yaldabaoth had designed, the capacity for deception and chaos that would make him one of the most dangerous beings in galactic history.

You might recognize him from the myths and legends that have filtered down to Earth across the ages, from stories told in the frozen North by peoples who retained fragmentary memories of cosmic events their ancestors had witnessed.

You might know him as Loki.

The Trickster. The Shapeshifter. The god who belonged to the Aesir yet worked against them, who was brother to Thor yet betrayed him, who brought both gifts and catastrophes to Asgard in equal measure.

The myths remembered him accurately in essence if not in detail. They understood his nature even if they forgot the cosmic scope of his actions. He was indeed chaos personified, deception made flesh, the insider-betrayer whose very existence would shake empires and reshape the course of galactic history.

Loki had been born into the game.

And nothing would ever be quite the same again.

11
THE JOINING OF TWO EMPIRES

C hrist Buri Anshar, wearing still the form of the Urmah war-
rior-king, wearing still the name Khan Anshar though millions of years had passed since he first took that title, had
been advancing and expanding his Asar Sirius Empire with relentless determination across vast stretches of the galaxy.

For ages beyond easy counting he had waged war against Ymir's
frozen legions and the corrupted Draco Empire, liberating world after world from their dominion, bringing human populations and allied species under the protection of the Asarrr banner, building civilization in territories that had known only predation and slavery.

Finally, inevitably, his expansion brought him to the borders of
territories he did not recognize, to worlds whose defenses bore
markings and tactical signatures unlike anything his intelligence
networks had reported from Draco or Ice Giant forces.

When he first encountered these worlds and observed their populations from a distance, when his scouts returned with reports of what
they had witnessed, Christ Buri made an assumption that seemed
entirely reasonable given his experience: he believed the humans
dwelling on these worlds had been captured and enslaved by the
dragons and reptilian beings who clearly held positions of authority
and military command throughout these territories.

After all, in every other region of the galaxy where humans
and dragons lived in proximity, the dragons were Draco—predators and slavers who used human populations as breeding stock
and Loosh generators. Why should this be different?

Operating on this mistaken belief, convinced he was liberating
enslaved humans from reptilian oppression, Khan Anshar ordered
his forces to attack these worlds with the full might of the Sirius
Empire's military apparatus.

Several major battles ensued—brutal confrontations where Sirius forces employing Urmah ferocity and Lycan savagery and Sasquatch strength crashed against defensive positions manned by

beings they assumed were enemies of humanity, never pausing to consider that the humans they claimed to be liberating were fighting alongside the dragons rather than cowering behind them.

The defenders fought with unexpected coordination and tactical brilliance, repelling the Sirius assaults with casualties on both sides that would have been catastrophic had these forces not been so vast that even heavy losses barely dented their overall strength.

After these initial engagements proved far more costly than Khan Anshar had anticipated, after it became clear that these were not hastily organized slave-world defenses but rather professional military forces operating with strategic depth and technological sophistication, the full Orion fleets were summoned from their staging areas throughout the Empire.

When they arrived, when the true scale of Orion's military might became visible to Sirius reconnaissance, Christ Buri found himself facing an opposing force unlike anything he had encountered in all his millions of years of warfare against Ymir and the Draco.

The Orion fleet was not merely large—it was vast beyond his previous experience, incorporating thousands of capital ships and tens of thousands of support vessels, manned by multiple species working in perfect coordination, wielding weapons that drew upon both advanced technology and cosmic principles his own forces had only begun to master.

In a display of overwhelming power calculated not to destroy but to demonstrate, to communicate through sheer scale that further aggression would be futile and suicidal, the Orion fleet maneuvered into positions that surrounded the Sirius expeditionary forces completely.

They could have annihilated Khan Anshar's armies in minutes. They chose instead to force him to stand down, to cease hostilities, to agree to a parlay rather than face mutual destruction.

Christ Buri, recognizing both the military reality before him and the possibility that he had made a catastrophic error in his initial assessment, agreed to meet with the Orion commanders under terms of temporary truce.

The meeting took place aboard a neutral station hastily established for this purpose, with both sides bringing their highest-ranking military officers and maintaining armed forces at ready positions should the negotiations collapse into renewed violence.

It was there, in that improvised diplomatic space, that Khan Anshar learned the truth he should have discovered before launching his attacks: he had not reached territories conquered by the Draco and their enslaved human populations, but rather had stumbled into Tiamat's kingdom—the Orion Empire, where humans and dragons and reptilians of various species lived as allies, fighting together against the Draco rather than serving them as slaves or suffering under them as victims.

The revelation struck him like a physical blow. He had attacked the very people he should have been allying with, had killed defenders who were on his side in the greater cosmic struggle, had nearly triggered a devastating war between the two largest powers opposing Ymir's expansion.

After explanations and apologies had been exchanged, after the military commanders had established that both empires were indeed fighting against common enemies rather than against each other, Khan Anshar was formally invited to attend a diplomatic meeting with the Orion Queens themselves—not as prisoner or supplicant, but as fellow ruler whose strength and purpose aligned with their own despite the unfortunate beginning to their relationship.

The journey to the Orion Throneworld took him through territories he had been preparing to "liberate," past worlds he now understood were already free, already thriving, already part of a

civilization as advanced and as dedicated to opposing the Demi-urge as his own Sirius Empire.

Upon reaching the Throneworld—that planet at the heart of the Orion Belt's central star, where Tiamat and Sabbaoth had established their Heaven and their seat of governance—Christ Buri Anshar was brought before the ruling council.

And there, after millions of years of separation, after ages spent fighting the same enemy from different fronts without knowing the other existed, Tiamat reunited with her Father from Heaven.

For though Christ wore Urmah flesh and bore the name Khan Anshar, though millions of years had transformed his appearance and his mortal identity, Tiamat recognized him instantly for what he truly was: Christ himself, the Savior, her grandfather through the divine lineage that connected her mother Pistis Sophia to God the Father and Sophia the Mother.

The reunion was joyous beyond measure, layered with relief that they had discovered each other before their forces had inflicted truly devastating casualties on each other, filled with wonder at how the game had arranged this meeting through what seemed like accident but was surely design.

They spoke at length about the war against Ymir and the Draco, comparing intelligence and strategic assessments, discovering that their separate campaigns had been more effective than either had realized because the enemy had been forced to fight on multiple fronts simultaneously. They discussed the possibility of coordination, of joint operations, of combining their strengths into a unified front that Yaldabaoth could not hope to overcome.

And as their conversation deepened, as political considerations merged with personal affection and strategic necessity aligned with familial bonds, they reached a decision that would reshape the galactic power structure entirely:

They would formalize their alliance through a political marriage, uniting the Sirius and Orion Empires not merely through treaty but through blood, creating a dynasty that would bind both realms together across generations to come.

Buri's son Bor was the eldest of his unwed sons—powerful in his own right, trained in warfare and statecraft, carrying Christ's divine bloodline through his father while possessing the Wingmaker heritage of his mother Kishar. He was an ideal candidate for such a strategic marriage, his status high enough to honor the alliance while his youth ensured the union would produce children who could inherit both empires.

To create the strongest possible offspring from this union, to ensure that the children born of this marriage would possess capabilities approaching the divine, Tiamat chose Nommu to be Bor's bride.

The logic was impeccable from a genetic and spiritual perspective: Nommu was not merely one of Tiamat's seven daughters but the most spiritually gifted among them, she who might be Sophia incarnate, she who carried the purest and most powerful Yin bloodline in the entire galaxy. United with Bor, who carried through his father the strongest Yang bloodline in existence, she would produce children of unprecedented potential.

What Tiamat did not know—what no one except Abraxas and Yaldabaoth knew—was that Nommu had already been corrupted by Abraxas's seduction, had already borne the hybrid Loki whose very existence was designed to undermine the Orion Empire from within.

The marriage was arranged and celebrated with appropriate cosmic grandeur, joining two empires and two bloodlines in ceremonies that lasted for weeks across multiple worlds, attended by representatives from every allied species and every constituent kingdom of both realms.

Bor and Nommu were wed according to rites both ancient and improvised, drawing upon traditions from Heaven itself and from the various cultures that had evolved across millions of years in the material frequencies. The union was blessed by Tiamat and Khan Anshar, by the seven Queens and the eleven Queensguard, by priests and mages and spiritual authorities representing every major tradition within the combined empires.

And soon after the marriage was consummated, soon after Bor's Yang seed met Nommu's Yin essence in the sacred act of creation, Nommu discovered she was pregnant once more.

The child gestated under the most favorable conditions imaginable, surrounded by cosmic forces aligned to ensure his healthy development, monitored by the greatest healers and seers available to either empire, blessed by powers both material and divine.

When the time came for birth, when Nommu brought forth the child who had been conceived from the union of the galaxy's two strongest bloodlines, a son emerged who would exceed even the most optimistic predictions of what such a hybrid might achieve.

His name was Ea—though many across countless worlds and ages would come to know him by another name, a title that would echo through mythologies and legends, that would inspire both worship and terror depending on who spoke it and in what context.

You know him today as Odin.

All-Father. Wanderer. God of wisdom and war, of poetry and death, of magic and kingship. The one-eyed seeker who sacrificed much to gain knowledge, who hung upon the world-tree to grasp the runes, who would one day rule Asgard and lead the Aesir through Ragnarok itself.

But in those early days, he was simply Ea—a child born from the convergence of cosmic forces, carrying within his hybrid ge-

netics the potential to become what Sophia had always intended the game to produce: a being capable of ending Poverty itself, of solving the riddle at the heart of creation, of discovering through lived experience why love exists and how abundance can be shared without creating new forms of lack.

As Bor was the son of Christ—the ultimate Yang/Soul/Assembly bloodline in all of existence, the divine masculine principle made flesh—and Nommu was the daughter of Tiamat and possibly the incarnation of Sophia herself—the ultimate Yin/Spirit/Life bloodline in the galaxy, the divine feminine principle walking among mortals—Odin was destined for greatness from the moment of his conception.

Destiny, fate, divine design—whatever term one prefers, all pointed toward this child as something unprecedented, something necessary, something that the entire cosmic drama had been building toward across millions of years and countless lifetimes.

As a baby, as an infant barely emerged from the womb, Odin quickly displayed capabilities that astonished even those who had expected remarkable things from him. He learned to talk within days of birth, forming words with clarity that should have required months or years to develop. He learned to walk almost immediately after achieving basic motor control, moving from helpless infant to ambulatory child in a span that defied normal developmental timelines.

His wit was incredible even in infancy, his understanding of complex concepts apparent to anyone who spoke with him, his ability to grasp cosmic principles evident in questions he asked before most children could speak their own names.

Observing this prodigy, recognizing in Odin the potential to become the key that would unlock the final mysteries of the game of Poverty, one of Tiamat's twelve Queensguard approached his

Queen with a request that surprised her despite all her millions of years of experience with the unexpected.

Heimdall was his name—possibly the most powerful of all the Queensguard, certainly the most unique in his origins. He had been created not merely by Tiamat alone but through the combined energies of the Dragon Queen and all seven of her divine daughters working together in unprecedented unity, weaving Yang and Yin in patterns never before attempted, producing a warrior whose capabilities exceeded even his eleven brothers.

He came before Tiamat and spoke with characteristic directness: "I wish to step down from my position as your Queensguard. I wish to dedicate the remainder of my existence to guarding Odin instead, to ensuring his survival and his development, to protecting him from all threats until he achieves the greatness for which he was born."

Tiamat, understanding the magnitude of what Heimdall was offering—his entire purpose, his role as her protector, the identity he had carried since his creation—considered his request carefully.

She recognized the wisdom in it. Odin would need protection beyond what normal security could provide. He would be targeted by Yaldabaoth and the Draco, by anyone who recognized his potential and sought to prevent him from achieving it. Having Heimdall as his dedicated guardian would increase his chances of survival dramatically.

But more than strategic calculation, Tiamat also recognized love in Heimdall's request—not romantic love but the fierce protective devotion that the greatest warriors sometimes feel toward those they have been called to defend, the willingness to subordinate all personal ambition and glory to the service of a cause greater than themselves.

"I grant your request," Tiamat declared. "You are released from your service to me, that you may serve my grandson and

protect the one who may finally end Poverty within the game. But to honor your sacrifice and your dedicated service across all these millions of years, I will not replace you. Henceforth I will have only eleven Queensguard instead of twelve, and the twelfth position will remain forever vacant as a testament to Heimdall who gave up his place to guard the chosen one."

And so it was established. Heimdall became Odin's personal protector from that moment forward, watching over the child as he grew, ensuring no harm came to him from any quarter, training him in arts both martial and mystical as he matured from prodigy infant into exceptional youth.

Odin remained on the Orion Throneworld during his childhood, learning at the feet of the greatest teachers the combined empires could provide. His mother Nommu instructed him in the mysteries of Yin energy and spiritual wisdom. His grandmother Tiamat taught him personally the secrets of becoming a creator god, sharing knowledge she had gained from Pistis Sophia and Sophia the Mother herself, guiding him through principles that only those with proper bloodline and proper preparation could safely access.

And during this time of education and growth, Odin quickly developed a friendship with his half-brother Loki—the other child of Nommu, only a few years older, already demonstrating his own remarkable abilities in magic and shapeshifting.

Neither Odin nor anyone else in the Orion court knew that Loki was Abraxas's son, that his very existence was part of Yaldabaoth's plan to corrupt the Empire from within. They saw only two exceptional youths, two brilliant young beings who complemented each other perfectly—Odin with his Yang mastery and Loki with his Yin fluidity, Odin with his straightforward pursuit of wisdom and Loki with his clever sideways approach to problems, Odin destined for kingship and Loki seemingly content to serve as advisor and companion.

Together, these two half-brothers attended the great mage schools of Orion, institutions of learning that had accumulated knowledge across millions of years, where the secrets of manipulating Soul and Spirit energies were taught to those deemed worthy and capable of handling such power responsibly.

Both Odin and Loki proved to be exceptional students, mastering disciplines that challenged most practitioners for lifetimes, advancing through levels of understanding with speed that astonished their instructors. They became masters of magic in their youth, wielding forces that elder mages struggled to control, demonstrating capabilities that suggested they would one day rank among the greatest cosmic manipulators ever to walk the material frequencies.

Yet despite their similar aptitude for magical arts, their education diverged in one crucial respect: only Odin received special private lessons from Tiamat herself on the ultimate mysteries of how to become a creator god, how to wield the combined Yang and Yin forces at the level required to shape reality itself, how to speak things into existence and alter the fundamental parameters of the game.

Loki, for all his brilliance and all his natural gifts, was not granted access to these highest teachings. Whether this was deliberate exclusion based on his questionable paternity—for rumors about Nommu's seduction by Abraxas had begun to circulate despite efforts to suppress them—or simply recognition that Odin's perfect bloodline gave him capabilities Loki could never match, the result was the same.

Odin learned to create. Loki learned to manipulate and deceive. And the seeds of future betrayal, though not yet sprouted, lay dormant in the soil of this differential treatment, waiting for the proper conditions to germinate into catastrophe.

The years of education passed swiftly in the way time passes for those engaged in meaningful work and genuine growth. Odin and

Loki matured from youths into young adults, from students into accomplished practitioners, from potential into actualized power.

It was during this period of relative peace and focused development that disturbing news reached the combined Sirius-Orion command: the Draco Empire had begun launching major attacks against the human worlds within the Pleiades—that star cluster where the largest concentration of Lyran refugees had settled after the Great Expansion, where humanity had rebuilt some measure of the golden age they had lost.

The attacks were systematic and brutal, clearly intended not merely to raid for slaves but to conquer and hold territory, to establish permanent Draco dominion over human populations who had thought themselves relatively safe in their distant refuge.

Christ Buri Anshar, hearing of this assault on human worlds, immediately rallied his forces to march into battle once more. This was precisely the kind of threat he had dedicated his entire incarnation to opposing—the predation of the strong upon the weak, the imposition of slavery and suffering on those who sought only to live in peace and abundance.

By this time, Odin and Loki were both fully grown and fully educated, possessing knowledge and power sufficient to contribute meaningfully to military campaigns rather than merely observing from safety. They had trained in combat as well as magic, had studied strategy as well as cosmic principles, had prepared themselves to take their place among the warriors and leaders who would shape galactic history.

They went along with Bor and Nommu, joining the armies of Christ Buri Anshar as he led a massive expeditionary force toward the Pleiades, determined to free the human worlds from Draco occupation and to demonstrate that the combined Sirius-Orion Empire would not tolerate such aggression against any human population regardless of how distant or isolated they might be.

The stage was set for a confrontation that would test Odin's capabilities for the first time in actual warfare, that would reveal whether the prophesied chosen one could live up to the destiny that had been proclaimed for him since birth.

And Loki would be there beside him, wearing the face of loyal brother and trusted companion, carrying within himself the corruption his father had planted, waiting for the moment when Yaldabaoth's plan would require him to reveal his true nature and betray everything Odin held sacred.

But that moment had not yet come.

For now, they rode together toward the Pleiades, two brothers united in purpose, unaware that their bond was built on foundations that would one day crumble into betrayal and catastrophe beyond imagining.

12

ODIN'S YOUTH AND THE PLEIADES TRAP

Before Christ Buri Anshar and his assembled forces departed from Orion to march toward the Pleiades, before the great military expedition was launched to liberate the human worlds suffering under renewed Draco assault, many years had passed in what appeared to be productive peace—years during which the two great empires that had nearly destroyed each other through mistaken conflict instead learned to share knowledge and strengthen each other through deliberate cooperation.

The Sirius forces had spent those years sharing their spiritual teachings with the Founder and Wingmaker races dwelling throughout Orion's territories, introducing perspectives and practices that had been developed across millions of years of separate evolution in different regions of the galaxy.

Before Khan Anshar's arrival, Orion had been a civilization heavily weighted toward the feminine principle—a Spirit-focused, Yin-dominant, Life-energy-based culture that reflected Tiamat's own essential nature as Dragon Queen and mistress of the receptive creative forces. They excelled at channeling chaos into manifestation, at receiving inspiration and making it real, at working with the fluid and adaptive aspects of cosmic power.

The Sirius Empire, by contrast, had developed as a more masculine-principle civilization—Soul-focused, Yang-dominant, Assembly-energy-based, reflecting Christ's nature as the ultimate ordering force, the divine logos made flesh. They excelled at structure and system, at imposing pattern on formless possibility, at wielding the directing and organizing aspects of cosmic power.

The exchange of knowledge between these complementary civilizations proved extraordinarily fruitful. Orion's Yin masters learned Yang disciplines they had previously struggled to grasp, while Sirius's Yang practitioners gained access to Yin techniques that had eluded their more structured approaches to cosmic manipulation.

During this period of cultural and spiritual exchange, Nommu and Bor—now formally known by his title as Bor Anu, taking the "An" prefix that marked him as master of Yang energy and legitimate heir to Khan Anshar's throne—spent much of their time among the courts of Orion, sharing the knowledge they had gained from their respective lineages.

Nommu taught what Sophia had taught her, what Tiamat had revealed to her, what she had learned through her own profound spiritual connection to the deepest mysteries of Yin. Bor Anu shared what Christ had taught his father, what Khan Anshar had passed to him, what he understood of Yang principles through both inherited wisdom and personal achievement.

Christ Buri Anshar himself traveled extensively throughout Orion during these years, visiting world after world, giving public teachings and private instruction, spreading understanding of cosmic principles to any who demonstrated readiness to receive such knowledge.

But his travels were motivated by more than mere educational generosity. He was also actively investigating and attempting to stop a disturbing trend he had noticed spreading through the Orion Empire like a slow poison—a corruption of values that stood in direct opposition to everything the game of Poverty was designed to teach.

The House of Abraxas had grown tremendously wealthy and powerful during the millions of years since its founder had returned from his training with Ymir. Through manipulation of markets and strategic control of key resources, through exploitation of legal loopholes and cultivation of political influence, Abraxas and his descendants had built an economic empire that rivaled some planetary governments in its reach and authority.

But wealth had not satisfied the hunger inherited from Yaldabaoth. Power had not filled the void at the center of Abraxas's

being. The more the House accumulated, the more it wanted, the more it needed, until want and need became indistinguishable from identity itself.

The House of Abraxas had begun harvesting and destroying entire worlds for profit, extracting resources with complete disregard for ecological sustainability or the wellbeing of populations dwelling on those planets. They would strip-mine worlds until nothing remained of value, would clearcut forests and drain oceans and poison atmospheres, leaving behind devastated husks where once had been living ecosystems.

And when the resources of a world had been fully extracted, when nothing remained to profitably harvest except the people themselves, the House of Abraxas had discovered an even darker source of revenue: they were slaughtering their slaves to farm their essence, their life-force, their Loosh—selling this psychic harvest to entities in the Draco Empire and even to certain corrupt factions within Orion itself who had learned to feed on suffering as Yaldabaoth's Archons fed.

The practice had become so widespread, so normalized within Abraxas's corporate structure, that managers overseeing these operations viewed it as simply another form of resource extraction, no different morally than mining ore or harvesting timber.

Christ Buri Anshar, discovering the full scope of this evil, launched a campaign to stop it through both legal and moral means. He visited worlds, documented the atrocities, gathered testimony from survivors, presented evidence before Orion's ruling councils in forums where truth could not be easily suppressed or dismissed.

He lobbied Tiamat directly, appealing to her as both political ally and family member, explaining how the House of Abraxas's practices violated every principle the Orion Empire claimed to uphold, how slavery and essence-harvesting were literally feed-

ing Yaldabaoth's power by generating exactly the kind of suffering the Demiurge required to grow stronger.

Tiamat, hearing the evidence and recognizing the corruption that had been allowed to fester within her own Empire while she focused on external threats, used her authority to push comprehensive reforms through Orion's legislative structures.

Laws were passed throughout the Empire that banned slavery in all its forms—not merely the crude chattel slavery of physical chains, but also the economic slavery of debt bondage, the political slavery of forced labor, the spiritual slavery of essence-harvesting. Strict regulations were established governing the extraction of planetary resources, limiting exploitation, requiring environmental restoration, mandating that local populations benefit from any development rather than merely suffering its costs.

These reforms struck directly at the heart of the House of Abraxas's business model, crippling their most profitable operations, forcing them to either adapt to ethical practices or cease activities entirely in territories where Orion law held authority.

The economic impact on Abraxas's empire was devastating. But equally significant was the strategic problem these reforms created: the House of Abraxas maintained vast private military forces—corporate armies originally assembled to protect their resource extraction operations and suppress slave revolts, soldiers and ships numbering in the hundreds of thousands.

Now these armies had nothing to do. The operations they had been created to defend were illegal. The populations they had been trained to control were freed. The worlds they had occupied were being evacuated and restored.

Idle armies are dangerous. Soldiers without purpose become restless, become susceptible to alternative employments, become available to anyone who can offer them action and glory and the sense of meaning that warfare provides.

But Abraxas, ever cunning, ever playing the long game his father Yaldabaoth had taught him, found a solution to his twin problems—how to restore his House's fortunes and how to employ his now-idle military forces.

During this same period, as Christ was campaigning for legal reforms and Tiamat was restructuring her Empire's economic foundation, the House of Abraxas was also quietly buying controlling interests in media centers throughout Orion—news networks, entertainment corporations, information distribution systems that shaped how millions of beings understood reality and their place within it.

Through these newly acquired media platforms, Abraxas began creating and promoting new forms of entertainment designed to distract Orion's populations from the spiritual life they had traditionally led. Where once the culture had emphasized meditation and cosmic understanding, growth toward creator-god status and alignment with divine principles, now the dominant messages encouraged materialism and sensual pleasure, competition for status and accumulation of possessions.

The shift was subtle at first, barely noticeable to those living through it. But across decades and centuries, the cumulative effect transformed Orion culture in ways that served Abraxas's purposes perfectly: populations focused on entertainment and consumption were far less likely to notice corruption in their midst, far less likely to organize resistance against exploitation, far more susceptible to manipulation through fear and desire.

And when the news came that the Draco Empire had launched massive assaults against the human worlds in the Pleiades and surrounding systems, when reports of atrocities and enslavement flooded through Abraxas-controlled media networks with carefully crafted emotional manipulation designed to inflame rather than inform, Abraxas found he could rally large portions of the Orion Empire to war with remarkable ease.

The populations wanted action, wanted spectacle, wanted to feel heroic and righteous. The media provided them with enemies to hate and causes to champion. And conveniently, the House of Abraxas had armies perfectly positioned to lead these newly militarized populations into glorious battle against the Draco threat.

What Tiamat and Sabbaoth did not know—what Christ Anshar suspected but could not yet prove—what even most members of the House of Abraxas did not fully understand—was that the entire situation had been orchestrated from the beginning.

The House of Abraxas had secretly pledged loyalty to Ymir, to Yaldabaoth wearing ice-giant form, to the Demiurge who had trained the House's founder in the darkest of archonic arts. Abraxas himself had traveled to Ymir's frozen domains periodically across the millions of years since his initial corruption, receiving updated instructions, coordinating strategy, ensuring his actions served the larger plan Yaldabaoth had designed.

And Ymir, working through his network of Archonic whispers and corrupted Draco servants, had been deliberately manipulating the Draco Empire into launching full-scale assaults on the Pleiades—not because those particular human worlds held special strategic value, but because attacking them would trigger exactly the response Yaldabaoth desired: the mobilization of both Sirius and Orion forces, the commitment of their best warriors and commanders to a distant campaign, the creation of opportunities for infiltration and sabotage and betrayal far from the oversight of Tiamat and Khan Anshar.

The trap was being prepared. The pieces were moving into position. And none of the intended victims yet understood how comprehensively they had been manipulated.

While these cosmic machinations were unfolding in political and economic spheres, while empires were being subtly redirected toward catastrophe through means their leaders could not perceive,

Odin and Loki were still living the lives of exceptional students among the greatest mages and creator gods in the Orion Empire.

They attended the most prestigious academies, studied under teachers whose knowledge spanned millions of years, mastered disciplines that challenged even naturally gifted practitioners. Their education was focused on preparing them for a specific role that held great honor within Orion culture: when creator gods went forth to seed new worlds throughout the cosmos, seven worthy and fully-trained practitioners were typically assembled into a working group known as Titans.

These Titans would travel together to uninhabited or primitive worlds, would coordinate their efforts to establish ecosystems and seed life, would remain until their created civilizations achieved sufficient stability to continue developing without constant divine oversight. Being chosen as a Titan was the highest achievement a Wingmaker could aspire to—recognition that one had mastered both Yang and Yin sufficiently to create balanced and sustainable realities.

Odin and Loki were both training specifically for this role, both working toward the moment when they would be deemed worthy to join a Titan working-group and venture forth as creator gods in their own right. And within their cohort of students, within the competitive environment of the most elite magical academy in the Empire, they stood at the very top of their class—surpassing even students who had studied for centuries longer, demonstrating capabilities that suggested they might one day rank among the greatest Titans who had ever lived.

It was during this period of intensive study and friendly rivalry that events began accelerating toward the crisis Yaldabaoth had been engineering.

Loki's father Abraxas—whom Loki knew only as a distant but influential figure, powerful in commerce and politics but rarely

personally involved in his son's life—sent word requesting a private meeting. The message was urgent, marked as highest priority, suggesting that something of tremendous importance needed to be discussed immediately.

Loki, curious and perhaps flattered that his father had finally chosen to acknowledge him directly, made arrangements to meet Abraxas in private, away from the academy, away from Odin and the other students, away from any witnesses who might observe or remember what was said.

What passed between them in that meeting, what revelations Abraxas shared with his son, what instructions or commands or manipulations were delivered—of this, no record exists except in Loki's memory and Abraxas's carefully maintained silence.

But we can infer much from what followed.

Meanwhile, Odin—hearing that his father Bor Anu would be leading a significant portion of the combined Sirius-Orion forces toward the Pleiades to fight the Draco, hearing of the great military expedition being assembled to liberate human worlds from renewed oppression—felt a powerful desire stirring within his young warrior's heart.

He wished to join the campaign. He wanted to fight alongside his father and prove himself in actual combat rather than merely training exercises. The Sirius Empire had been experiencing a period of relative peace since allying with Orion, and Odin had grown up hearing stories of his father's and grandfather's great victories but had never personally experienced the glory of battle.

He wanted war accolades like those Bor Anu carried from previous campaigns. He wanted to earn recognition as a warrior in his own right rather than merely being known as the grandson of Christ and Tiamat, the inheritor of the greatest bloodlines. He wanted to discover whether he was truly the chosen one proph-

ecies proclaimed him to be, or merely a child of privilege whose supposed destiny existed only in others' hopes and expectations.

And he had trained for precisely this kind of challenge. Heimdall had spent years teaching him combat arts both physical and mystical, had pushed him harder than any other student was pushed, had demanded perfection because anything less might mean death when real enemies attacked. Odin had become a formidable warrior—not merely competent but exceptional, not merely trained but genuinely gifted in the application of force and strategy.

He believed himself ready. And perhaps he was ready, in purely martial terms. But readiness for combat and readiness for the cosmic game Yaldabaoth was playing were entirely different matters, and in the latter Odin remained dangerously naive.

It is worth noting here the nature of Heimdall's own journey, for it illuminates much about the forces gathering around Odin and the preparation that had been invested in his survival and success.

When Heimdall had first pledged to serve as Odin's personal bodyguard upon the child's birth, when he had offered to step down from his position among Tiamat's Queensguard to devote himself entirely to protecting the prophesied chosen one, Christ Buri Anshar had delivered an unexpected response:

"No," Khan Anshar had said. "Not yet. Not until you complete a specific training that will make you worthy of this sacred duty."

He had required Heimdall to master the discipline of the An-Ki-Els—a term many across various worlds and ages would come to know through corruption and translation as "Archangels," though this rendered only partially the full meaning of the original name.

An-Ki-Els. The Yang-Life Warriors. The Divine-Earth Mighty Ones. The angelic beings who served as Christ's primary military com-

manders and most trusted instruments of divine will in the material realm.

They were masters of Soul and Yang and Assembly energies—the masculine principles of order and structure carried to their highest expression. But they were also trained Wingmakers and Founders, possessing the spiritual power and creative authority that came from achieving full enlightenment and direct connection to Source. They combined these dual masteries into a synthesis that made them devastatingly effective both in combat and in spiritual ministry.

Like the Paladin archetype known in various fantasy traditions—like the holy warriors who combined priestly authority with martial prowess, who could heal with one hand while smiting with the other—the An-Ki-Els were super-powered warrior-priests capable of performing genuine miracles both in battle and among civilian populations.

They could speak blessings that actually manifested as tangible protection, could pronounce judgments that materially affected the judged, could wield light itself as a weapon against forces of darkness. They were, in the most literal sense, angels—messengers and enforcers of divine will, beings who walked the material frequencies while channeling power from the highest realms.

Most of the An-Ki-Els were humans who had earned their wings as Wingmakers—individuals who had achieved full enlightenment through study and discipline, who had been granted the physical manifestation of wings as a mark of their ascension, who had then chosen to dedicate their elevated capabilities to Christ's service rather than pursuing purely personal spiritual advancement.

Heimdall, however, presented a unique case. He was a shape-shifting dragon who usually took the Aryan form of a Wingmaker when manifesting in humanoid shape—tall, fair-haired, blue-eyed, possessing the distinctive markers of the dragon-human

hybrid race Tiamat had created. Because of his dragon heritage, because he had been born from the combined energies of Tiamat and her seven daughters who were themselves the purest sources of Yin in the galaxy, Heimdall connected naturally and powerfully with Spirit and Life energies.

But he had never been strong with Soul and Yang and Assembly energies. The masculine principle that came so naturally to Christ's forces, that Buri Anshar wielded with effortless authority, remained difficult for Heimdall to channel effectively. He could use Yang force, certainly—his combat capabilities proved that. But mastery eluded him, fluency escaped his grasp.

To become an An-Ki-El, he would need to overcome this limitation entirely, would need to achieve balance between Yin and Yang that matched Odin's own prophesied capabilities.

Heimdall had trained for many years with the An-Ki-Els, studying their techniques, learning their disciplines, struggling to master energies that resisted his dragon nature. He had even received direct personal instruction from Christ Buri Anshar himself—Khan Anshar taking time from his campaigns and his governance to personally train this devoted warrior who sought to guard Christ's prophesied grandson.

Through dedication that bordered on obsession, through willingness to endure failure after failure until success finally emerged, through the particular kind of love that manifests as absolute commitment to protecting another, Heimdall eventually succeeded in his training.

He became a master of both Yin and Yang, of both Spirit and Soul, of both Life and Assembly. He achieved the balance required of an An-Ki-El, earned the rank and authority that came with completing their legendary training program, proved himself worthy to stand among Christ's mightiest warriors as an equal rather than merely an aspirant.

He had become not just a mighty warrior but a complete warrior—one who could fight on any battlefield, wield any energy, counter any threat, protect his charge from enemies both material and spiritual.

Now, with war approaching, with the Pleiades campaign about to begin, Heimdall and Odin would finally go forth together to claim glory against the Draco, to test in actual combat what had been learned through years of preparation, to discover whether prophecy and training could withstand the chaos of real warfare.

Together they marched toward the Pleiades, mentor and student, guardian and ward, An-Ki-El and prophesied chosen one.

But the Draco—or rather, Yaldabaoth working through the Draco, Abraxas working through his House's infiltration of Orion's military structure, Loki working through whatever instructions his father had delivered in their private meeting—had other plans in mind.

Plans that did not involve Odin covering himself in glory.

Plans that instead involved his capture, his corruption, or his death.

The trap was ready. The innocent were marching into it willingly, eagerly, believing they fought for liberation when they were actually serving the Demiurge's design.

ABOUT THE AUTHOR

FOR 50 YEARS, ONE MAN AWAKENED HUNDREDS OF THOUSANDS TO HIDDEN TRUTHS

Dr. Delbert Blair (1934-2016) wasn't just another metaphysics teacher—he was an engineer, research scientist, and historian who spent five decades exposing what mainstream institutions won't touch: the suppressed science of meditation, the dangers of EMF radiation before it was fashionable, the real history of Egypt's 18th Dynasty, and the metaphysical principles that govern reality itself.

From 1973 until his passing in 2016, Dr. Blair directed The Meta-Center in Chicago, lectured in over 200 forums worldwide, and built a community of truth-seekers who refused to accept surface-level explanations. His teachings combined rigorous scientific methodology with ancient wisdom—covering everything from cellular health dangers to sexual energy dynamics, from dual solar systems to the practical application of higher consciousness.

The lineage continues. Before closing The Meta-Center in 2015 due to health challenges, Dr. Blair personally authorized his protégé—Tony 'The Vortex' Blair, whom he called his "God Son"—to carry forward the work.

Tony studied directly under Dr. Blair for years, absorbing the teachings, methodology, and mission that awakened a global community.

"There is no Truth until you decide what Truth is."

— Dr. Delbert Blair

In August 2015, with Dr. Blair's blessing, Tony reopened The Meta-Center as the only authorized source for Dr. Blair's teachings and materials. The work continues. The truth remains uncompromising. The community grows stronger.

This isn't just a memorial—it's a living mission for Humanity. Dr. Blair taught that we must be participants in the Great Work, not mere observers. His legacy lives through every student who raises their vibration, questions manufactured realities, and seeks genuine innerstanding.

TRAINED BY DR. DELBERT BLAIR.
TRUSTED TO CONTINUE A 50-YEAR LEGACY.

For two decades, Tony 'The Vortex' Blair has done what most people can't: see through the engineered narratives that shape your perception of reality.

While the masses consume carefully crafted stories from media, institutions, and intelligence operations, Tony decodes the behavioral tradecraft behind them. As a specialist in psychological operations, strategic influence, and counter-disinformation, he teaches you to recognize when you're being manipulated—and how to break free.

This isn't theoretical. Tony's expertise spans intelligence analysis, behavioral psychology, and strategic deception. He's advised professionals navigating propaganda environments, consulted on counter-narrative strategies, and helped individuals reclaim their cognitive autonomy in a world designed to control their thoughts.

But his deepest credential comes from the late Dr. Delbert Blair himself—the legendary metaphysics teacher and research scientist who awakened hundreds of thousands over 50 years. Dr. Blair didn't just teach Tony; he trusted him as one of the few minds worthy of bouncing ideas off. Before his passing, Dr. Blair personally authorized Tony to continue The Meta-Center's mission, calling him his "God Son" and chosen successor.Tony's approach combines multiple disciplines most "experts" never touch:

- **Behavioral Tradecraft & Psychological Operations:** Understanding how influence works at the neurological level—so you can spot it in real-time and in some cases advance your own mental processes

- **Engineering & Systems Thinking:** Deconstructing complex architectures the way an engineer reverse-engineers technology

- **Herbal Medicine & Bioengineering:** Studying natural healing systems since age 11, recognizing that physical sovereignty supports cognitive sovereignty

- **High-Altitude Research and Engineering:** Launching stratospheric balloons to 30+ miles above Earth, gathering atmospheric data beyond institutional control

Where others specialize narrowly, Tony connects dots across domains—because manipulation operates across all systems: biological, psychological, informational, technological.

The mission remains what Dr. Blair taught: participate in the Great Work of Humanity, don't just observe. Tony continues that legacy through teaching, writing, and providing frameworks for ethical influence—showing you how the game is played so you can stop being played.

This is advanced knowledge for serious students. If you're content with surface-level explanations, this isn't for you. But if you're ready to see the machinery behind manufactured consensus, Tony will show you where to look and how to protect and strengthen your inner-self.